Anya Heise-von der Lippe
Writing Romantic Climate Change

Literary Ecologies | Volume 3

Editorial

In the age of the Anthropocene and in the face of climate crisis, societal perceptions of the environment, the way it is talked about, and how it is represented aesthetically are changing. In Literary Studies, scholars increasingly focus on the relationship between humans and nature, on how the latter manifests itself as an important element in literature, and on how literature influences the ecological discourse of its time.
Building on ongoing discussions in Ecocriticism, the **Literary Ecologies** series examines the interplay between literature and the environment. It aims to illuminate perceptions and conceptions of the environment from ancient times to the present day as well as the tense relationship between humans and nature in different forms of literary expression. Special interest is also given to the complex relationship between human and non-human actors, which is increasingly gaining importance in debates about climate change, species extinction, and other forms of environmental degradation.
The series is not limited to analyses of fictional literature, but also integrates investigations of ecological narratives in non-fiction as well as theoretical discussions relevant to Ecocriticism itself. Possible topics and research fields in the series include Plant and Animal Studies, Petro- and Ecocriticism, Climate Fiction, Nature Writing, Ecopoetry and Literatures of the Anthropocene.

Anya Heise-von der Lippe teaches English literature and culture at Eberhard Karls Universität Tübingen. She holds an MA from Technische Universität Berlin and a PhD in English literature from Eberhard Karls Universität Tübingen, as well as a German post-doctoral degree (Habilitation) from Universität Potsdam. Her current research focuses on climate literature and the critical posthumanities.

Anya Heise-von der Lippe

Writing Romantic Climate Change

Gendered Poetics and Critical Legacies in the Anthropocene

[transcript]

Financial support for this publication was provided by the Equality Committee (Gleichstellungskommission) of the Faculty of Humanities at the University of Tübingen.

Bibliographic information published by the Deutsche Nationalbibliothek
The Deutsche Nationalbibliothek lists this publication in the Deutsche Nationalbibliografie; detailed bibliographic data are available in the Internet at https://dnb.dnb.de/

Cover layout: Maria Arndt, Bielefeld
Cover illustration: Tammo Möller
Printed by: Majuskel Medienproduktion GmbH, Wetzlar
https://doi.org/10.14361/9783839472750
Print-ISBN: 978-3-8376-7275-6
PDF-ISBN: 978-3-8394-7275-0
ISSN of series: 2941-4210
eISSN of series: 2941-4229

Printed on permanent acid-free text paper.

This one is for the next generation.

Contents

Acknowledgements

Almost 100 years ago, Virginia Woolf opened her seminal discussion of the conditions of women's writing in *A Room of One's Own* (1929) with "an opinion upon one minor point – a woman must have money and a room of her own if she is to write".[1] Woolf's argument concerns the writing of fiction, but I would argue that it, fortunately or unfortunately, also holds for any other kind of extended writing process. Nor has it lost any of its merit. Without the necessities of time – which can not be spent in other kinds of labor – or a space to write in that is free of interruptions, the probability of any extensive writing is almost nonexistent. I have explored some ideas around the necessity of an academic politics of care that would allow those who do care work and those who need care better chances of participation elsewhere and I do not wish to rehash them here.[2] Suffice it to say that the engagement with women writing in the early nineteenth century alongside critical discussions that still often dismiss their work based on their social position and the circumstances of their writing has made it abundantly clear to me that much critical groundwork still needs to be done around the situatedness of women's writing, especially in contexts where seemingly objective standards are applied without questioning the frequently gendered or otherwise marginalizing assumptions they entail.

Large parts of this book were written during a global pandemic which made travel for research purposes impossible and stranded many of us at home for months or years on end. This lack of access shaped the project in unforeseen ways, as it limited the canon of texts I was able to include to already published works. As a disabled academic living in the Global North, I had the privilege of being able to work from home – using the former pantry as my office space –, a mostly stable internet connection and the resources to buy the books that were not already accessible to me via online repositories. I also had the opportunity to attend online book launches, talks, conferences and symposia in faraway places like California, Australia or Canada, which would have been largely inaccessible to me before the pandemic and which

1 Woolf, *A Room of One's Own*, 5–6.

2 See Heise-von der Lippe, "Cripping the Neoliberal University – We need a Politics of Care".

have now, unfortunately, moved back to in-person attendance. There are, consequently, a number of people whose ideas – shared graciously with the world – influenced my own research processes and helped me develop my arguments, but whom I have never actually met in person. Wherever possible, I have attempted to find their published work and cited it throughout this book to acknowledge these contributions.

Closer to home, I wish to express my sincere gratitude to Russell West-Pavlov at the University of Tübingen for his continued support and countless helpful suggestions. I also wish to thank my reviewers, Molly Brown at the University of Pretoria, South Africa and Dirk Wiemann at the university of Potsdam for their willingness to review my project. I am very grateful to Anja Schwarz at the University of Potsdam, for taking on the task of organizing everything and for providing such excellent guidance throughout the process. The Habilitation was a big step for me, and I could not have wished for better circumstances or, indeed, a better experience. I also wish to thank my amazing students, with whom I discussed Climate Poetry, Fungi and Romantic Apocalypses, my colleagues in Tübingen and my network of academic friends around the globe who encouraged me from afar while I completed yet another big writing project. Last but certainly not least, I wish to thank my family for their continued support of my academic endeavors, especially Heiner von der Lippe, who went on long graveyard walks with me and who also took the mushroom picture in chapter 4.

As I prepare this manuscript for publication, the world continues to face multiple entangled crises that complicate each other, and I am truly grateful to the people who are actively working on keeping hope alive and on pushing against lethargy and climate doomerism. I hope that in showing some of the ways in which gendered assumptions complicate Romanthropocene connections and the ways in which these connections have shaped contemporary discourses around such fundamental terms as "the human" and "nature", I can at least contribute something to the understanding of certain aspects of the ongoing climate crisis that might help in imagining better ways of coping with it in the future.

Berlin, January 2024

Introduction

There was a time! that time the Muse bewails,
When Sunny-Hill enjoy'd refreshing gales;
When Flora sported in its fragrant bow'rs,
And strew'd with lib'ral hand her sweetest flow'rs!
Now sable vapours, pregnant with disease,
Clog the light pinions of the southern breeze;
Each verdant plant assumes a dusky hue,
And sooty atoms taint the morning dew.
No more the lily rears her spotless head,
Health, verdure, beauty, fragrance, all are fled:
Sulphureous clouds deform the rising day,
Nor own the pow'r of Sol's meridian ray;
While sickly damps, from Aire's polluted stream,
Quench the pure radiance of his parting beam.
These are thy triumphs, Commerce! – these thy spoils!
Yet sordid mortals glory in their toils,
Spurn the pure joys which simple Nature yields,
Her breezy hills, dark groves, and verdant fields,
With cold indiff'rence, view her blooming charms,
And give youth, ease, and health to thy enfeebling arms.[1]

Looking back from a twenty-first century European vantage point, Maria Logan's "Verses on Hearing That an Airy and Pleasant Situation, near a Populous and Commercial Town, Was Surrounded with New Buildings" first published in *Poems on Several Occasions* (1793) presents itself as an early-anthropocene climate poem. The poem describes the triumph of "Commerce" over the joys of "simple Nature" as "Sunny-Hill" is swallowed by the spread of urbanization. Rhyming couplets juxtapose the image of a personified "Flora" strewing her "sweetest flowers" and the "refreshing

1 Logan, "Verses on Hearing That an Airy and Pleasant Situation, near a Populous and Commercial Town Was Surrounded with New Buildings."

gales" and other "joys" of "simple Nature" with a scenery of soot, pollution and "sulphureous clouds", presented as a bitter argument (underlined by two exclamation marks) directed at "Commerce!", to "enjoy thy spoils!"

"Commerce", personified as the direct addressee of the poem, is seemingly detached from human agency, allowing the speaker to rail against an abstract concept, while the "sordid mortals glory[ing] in their toils" are described as misled and, in the end, doomed to suffer too, as they "give youth, ease, and health to [the] enfeebling arms" of "Commerce". This anti-commercial stance draws on the common Romantic register of juxtaposing the "too-muchness" of large-scale commercial endeavors as a form of exploitation with other, often rural, forms of work, which are framed as useful and beneficial to society. In early anthropocene texts, this is often supported by juxtapositions between country and city or agricultural and industrial work, where the former is framed as picturesque and in harmony with "nature" and other human beings, and the latter is condemned as exploitative, destructive and detrimental to people's health.[2] While it would be an oversimplification to read the world of work and trade as entirely male-connoted in women's poetry,[3] women writers of the Romantic period often employ distancing mechanisms, which allow them to take an anti-commercial position from an observer's perspective – as Logan's speaker does in the example quoted above and as I will show for other texts throughout this project.

In the time frame of Logan's poem, the anthropocene is already in full swing: the ephemeral "Airy and Pleasant Situation" is now "Surrounded with New Buildings", permanent in their solidity. The destruction of "nature" has been accomplished and the prevailing tone is reminiscent, rather than suggesting that it might be possible to return to the previous, idyllic state. From this retrospective point of view, the descriptions of "simple Nature" seem somewhat formulaic (still rooted in the orderly imagery of eighteenth-century Augustan poetry) in comparison to the more contemporary, vivid images used to describe the damage of urbanization and – although this is less prominent in the text – industrialization.[4]

2 Susanna Blamire's "Stockelwath" is an example of this juxtaposition of the "dark clouds of curling smoke" in industrial towns and the idyllic view of a country village and its neat apple orchards and herb gardens. The poem's speaker frames this view as a holiday associated with health, rather than showing the work associated with country life (Blamire, "Stocklewath; or, The Cumbrian Village," 115.).

3 See for instance Mary Elizabeth Robinson's "A London Summer Morning" (1794), which describes women engaged in various occupations alongside men: "The ruddy housemaid twirls the busy mop, / Annoying the smart 'prentice, or neat girl / Tripping with bandbox lightly." (Robinson, "A London Summer Morning," 183.)

4 The "Aire's polluted stream", mentioned in line 13, references a river running through Leeds, one of the centers of the woolen industry in the eighteenth century and a rapidly growing hub of the industrial revolution in Northern England.

Perhaps the most striking feature of Logan's poem from a twenty-first-century perspective aware of anthropocene discourses is the pervasiveness of its theme, which continues to resonate with current discussions of anthropogenic climate change. As Elaine Gan, Anna Tsing, Heather Swanson and Nils Bubandt argue in their Introduction to *Arts of Living on a Damaged Planet* (2017), "The winds of the Anthropocene carry ghosts – the vestiges and signs of past ways of life still charged in the present."[5] For the authors these ghosts are reminders of the traces humanity's past actions have left on landscapes and environments. Their aim is not to romanticize past landscapes, but rather to show

> how to pay better attention to the overlaid arrangements of human and nonhuman living spaces, which we call 'landscapes'. Our hope is that such attention will allow us to stand up to the constant barrage of messages asking us to forget – that is, to allow a few private owners and public officials with their eyes focused on short-term gains to pretend that environmental devastation does not exist.[6]

This argument against "short-term gains" sounds remarkably similar to Logan's poetic critique of "Commerce", connecting anti-capitalist arguments across the anthropocene. The poem's invocation of "simple Nature" is, however, rooted in a romanticized past that, even at the time of writing, was an aesthetic construct. What connects the two texts is, thus, not their specific stance on damaged "nature", but rather the fact that as texts engaging with damaged environments in the anthropocene they present an opportunity to reflect on human agency as a haunting presence across multiple temporalities. To read Logan's poem as an anthropocene text thus requires the additional critical lens of an anthropocene poetics that does not fall into the trap of romaniticizing an untouched state of "nature" that never existed and that, consequently, there is no possibility to return to. Instead, an anthropocene perspective might read the poem as an invitation to reflect the ramifications of past and ongoing anthropogenic changes and traces on the landscape, as well as a specifically localized perspective from which such traces might be described.

As I will argue in more detail below, the connection between Romantic literature engaging with the destruction of the non-human natural world and current perspectives on ecological destruction is a complicated one, as it is also based on an understanding of late-eighteenth- and early-nineteenth-century texts from a current ecocritical lens. As Richard Kerridge argues, ecocritical readings approach a text "from a very specific present" as "the sense of emergency with which ecocritics read belongs peculiarly to our own time."[7] While I agree with Kerridge's under-

5 Gan et al., "Haunted Landscapes of the Anthropocene," G1.

6 Gan et al., G1.

7 Kerridge, "An Ecocritic's *Macbeth*," 194.

standing of ecocriticism as a concern driven by current issues and discourses, I also argue that an ecocritical line of questioning may reach back into early anthropocene texts, offering openings for anthropocene readings. Moreover, for Romantic ecocriticism specifically, there has been a movement of ideas in the opposite direction, that is from Romantic poetry and poetics into ecocritical responses to Romantic writing, most prominently, as Jeremy Davies argues, into early Romantic ecocriticism (Davies cites Jonathan Bate's work specifically) in the 1990s, which is based on the common view that "the origins of environmentalism can be found in Romantic attitudes towards nature."[8] These critics were, as Davies argues, "glad to see themselves as the inheritors of Romantic thought" and "read Romanticism with rather than against the grain."[9] This critical continuity also extended to a limited focus of their readings on the canonical male Romantic poets, further anchoring a specific understanding of Romantic literature in Romantic ecocriticism.

While recent ecocritical texts have extended their scope beyond the "Big Six", theorizations of ecological thought in the Romantics were, until very recently, based largely on a relatively small group of canonized writers, leading to a narrow understanding of the ways in which "the Romantics" engaged with "nature". Davies cites Bate's understanding of Romantic ecocriticism from *The Song of the Earth* (2000):

> A central question in environmental ethics is whether to regard humankind as part of nature or apart from nature. It is the task of literary ecocriticism to address a local version of that question: what is the place of creative imagining and writing in the complex set of relationships between humankind and environment, between mind and world?[10]

It is not so much the line of questioning as the scope and underlying assumptions of the argument that need to be addressed in more detail, as it seems to use terms like "humankind", "mind" and "world" in a universalizing manner that does not only draw on Romantic ideas of "the human" as an all-encompassing subject position that does not question its exclusion of "Others", but also further romanticizes the canonized Romantics' specific understanding of the relationship between the poet's mind and "nature". As I argue, this positionality, which recent critical discourses have largely inherited from Romanticism, does not only continue to narrow their perspectives on a specific Romantic canon, but it has also further cemented an understanding of what it means to engage with "nature" that excludes other perspectives. I borrow the term "feedback loop" from climate discourse to describe these

8 Davies, "Romantic Ecocriticism," 3.

9 Davies, 3.

10 Bate cited in Davies, 3.

historical connections and their consequences. My argument is focused on broadening the critical lens by tracing continuities and connections going in two different historical directions, that is, between the Romantics' engagements with a sense of climate change and climate catastrophe at the beginning of the anthropocene, the critical framing of Romantic engagements with "nature" by Romantic and ecocritical critics in the twentieth century, and, in a further step, an understanding of the ways in which these discourses resonate with and resurface as romanticizations of "nature" as an externalized entity separate from "the human" in twenty-first-century discourses around climate crisis.[11]

From a current point of view influenced by climate crisis, I am particularly interested in the ways in which Romantic women writers' early engagements with climate change and climate catastrophe draw attention to a sense of urgency, that is their focus on specific kinds of situated knowledge and experiences that resonate with current socio-economic or cultural shifts and push against the Romantic urge to separate an autonomous self from the non-human natural world. In this context, I read Logan's poem as a response to changes that, while already perceptible at the time, had yet to be framed critically as anthropogenic climate change. The poem describes these changes from a very specific perspective and with specific poetic strategies, that I will present in a brief exemplary reading here.

Published in 1793, Logan's poem draws on an Augustan, perhaps more than a Romantic, conceptualization of "nature"[12] that, nevertheless, prefigures what might retrospectively be understood as a Romantic emotional engagement with a changing world. Most importantly for my argument, the poem presents an aestheticized stance against the anthropogenic impact on (local) climate that makes it an interesting starting point for this project: The opening line is interrupted by an exclamation mark, highlighting a sense of reminiscence of a previous time now lost to ecological destruction: "There was a time!" – but that time is already past. The "muse be-

11 An in-depth critical discussion of the constructedness and exclusiveness of the concept of "the human" that Romanticism inherits from Enlightenment theorists like Locke and Rousseau goes beyond the scope of this text. While I use the terms "human" and "humanity" to discuss Romantic discourses of the position of human beings towards and within the non-human natural world, it should be clarified that this discursive position is essentially anthropocentric, Eurocentric and otherwise exclusionary and has been questioned from various critical positions (including feminism and critical theorizations of constructions of race, as well as contemporary critical posthumanism). For a more detailed discussion of the inception and legal ramifications of the concept of "the human" see, for instance, Perry, *Vexy Thing – On Gender and Liberation*.

12 I use quotation marks to signal the concept of "nature", specifically as the term is used in Romantic writing and criticism (see also Day, *Romanticism*, 34ff.). The general setting and entities outside of a human body that do not carry this specific set of connotations are referred to as "non-human nature/the natural world".

wail[ing]" this lost past underlines the poem's aesthetic function in verbalizing the memory as well as the current destruction within an aestheticized poetic context. It also functions as an externalization of the speaker's emotions. It is, after all, not the speaker who is "bewail[ing]" this past aesthetic experience but a more generalized "muse" who is called on as an aesthetic arbiter and a poetic voice to rail against the extent of the destruction. The emotional reaction to the changed climate is expressed in the poem's language, direct address and punctuation, but it is curiously detached from the speaking subject.

In its indirect bemoaning of the lost natural idyll of the past, the covert speaker's position is also both temporally and spatially detached from other humans – the "sordid mortals" – and takes up a distanced, judging position instead. Emotions are, thus, expressed either by proxy (the muse) or in the poem's structure. The juxtaposition of ordered "nature" and destructive unruliness is underlined by the iambic meter, which remains regular in the lines describing "nature" but shows some irregularities when the poem is most forceful in tone and argument. Apart from the pauses or emphases associated with punctuation, two of these interruptions of the poem's regular rhythmic flow occur at the beginning of lines 6 and 17, drawing attention to the sudden emphasis on the verbs "clog" and "spurn" at the beginning of these lines. The emphasis on "Clog" serves to underline the speaker's outrage at the pollution.[13] "Spurn" carries a similar note of anger at the "sordid mortals" who are no longer interested in "nature". The text, however, also warns that this disconnect will ultimately be detrimental to them, establishing a connection between the "mortals" and "nature" by drawing attention to their mortality.

While the text itself does not present any gender markers, its complicated construction of a detached speaker position is an example of the complex ways in which women writers in the Romantic period engage with climate change and climate catastrophe. As I will argue, the textual examples I discuss draw attention to the problem of engaging with the non-human natural world from a perspective that is excluded from the experience of the autonomous self and its relationship with "nature", based on an understanding of poetic genius as male-connoted in Romantic aesthetics. As I argue, climate change and climate catastrophe trigger a sense of urgency and acceleration that foregrounds the necessity of developing a subject position beyond the autonomous self from which to describe a sense of change which also affected the relationship between the experiencing subject and the non-human natural world. While strategies differ individually, as well as in terms of genre and theme, there are some commonalities between the texts I discuss

13 While "clog" is used in direct connection to the "vapours" now clogging the "southern breeze", the addition of "pinion" with its double meaning of birds' wings and machine parts might be read as a reference to the original practice of sabotage, but this may well be an anachronistic reading of a poem published in 1793.

in the following chapters, including attempts at constructing distance, attempts at anchoring a situated subject position in embodied experience, and narrative strategies that split reactions to climate catastrophe across different characters.

My argument is based on an understanding of positionality that comes from feminist epistemology, most prominently the work of Donna Haraway on situated knowledges.[14] While the term "positionality" usually refers to an understanding of the fact that intersecting social and political contexts will have an impact on a person's identity – for instance in terms of gender, racialization, class or disability – feminist epistemology argues that knowledge and knowledge production are shaped by positionalities and the power structures they inhabit.[15] These structures are usually invisible in stable systems, but may become visible in moments of change or crisis and in the cultural productions that engage with them. I read Romantic responses to climate change and climate catastrophe as an opportunity to highlight shifting engagements with positionality in this sense.

Rather than propose a rigid gendered difference solely on the basis of the (perceived) gender of the author of a specific text,[16] I would like to suggest that the socio-political and cultural inequalities faced by women writers of the Romantic period led to a heightened awareness of exclusion from cultural and aesthetic discourses that tend to emerge in their writing as a troubling of the dominant subject position. This is especially visible in texts concerned with circumstances or events that might undermine the highly contested relationship between "man" and "nature", such as poetic or narrative engagements with climate change and climate catastrophe. While this has farther-reaching consequences for the ways in which women write "nature",[17] my focus here is on texts that engage with major changes to the concept of "nature" in the context of climate crisis and the shifts in positionality they trigger.[18]

Logan's poem quoted above presents an early example of writing in response to climate change with a distanced speaker whose positionality can be constructed

---

14 Haraway, "Situated Knowledges: The Science Question in Feminism and the Privilege of Partial Perspective."

15 see Grasswick, "Climate Change Science and Responsible Trust."

16 Here, I follow Mellor's assertion that "the terms 'masculine' and 'feminine' Romanticism are ideological or socially constructed terms, not biological, sex-based terms" (Mellor, "Thoughts on Romanticism and Gender," 345.).

17 Mellor, for instance argues that women writers of the Romantic period "conceive of [Nature] as a woman with her own needs as well as an infinite variety of resources". In their view, "evolution will come about not through the mastery of Nature, but rather through co-operation with Nature" (Mellor, 345.). For a more recent ecofeminist discussion see Stevens, Tait, and Varney, *Feminist Ecologies*.

18 My argument is, thus, not necessarily focused on the non-human natural world that emerges in these texts, but on the ways in which the poetic or narrative subject positions itself towards it.

as the absence of an authoritative subject position that might speak about or on behalf of the "simple Nature" under attack in the text. A brief comparison to William Wordsworth's sonnet "The world is too much with us", written circa 1802 and published in 1807 – a decade later than Logan's poem but focused on a very similar topic –, shows the difference in speaker positionality in the two poems. There is considerable overlap in the stance the two poems take against "Commerce" – framed as "getting and spending" and further highlighted by a shift from iambic to dactylic meter in Wordsworth's poem.[19] They also both lament the disconnect between humans and "nature" that they frame as "sordid" and associate with a loss of "our powers" (Wordsworth) or "youth, ease, and health" (Logan).[20] The most prominent difference between the two texts hinges on the question of who is included in the dominant social group the poem criticizes and, thus, implicated in having lost connection to "nature" in Wordsworth's sonnet. While "The world is too much with us" seems to confidently encompass a unified group that might be identified as humanity in general, Logan's poem frames this group as "sordid mortals", not including the speaker in this group. This distinction in positionality has an impact on the authority and confidence of the speaker that is also expressed in the phrase "our powers" in Wordsworth's texts, as the speaker is included in the group of people who have turned their back on "nature". The speaker, thus, feels able to make an authoritative statement about their emotional disconnect:

> The world is too much with *us*; late and soon,
> Getting and spending, *we* lay waste *our* powers; –
> Little *we* see in Nature that is *ours*;
> *We* have given *our* hearts away, a sordid boon![21]

We/our pronouns occur seven times in the first four lines, leaving no doubt about the speaker's firm inclusion of their own position within the group of people that Logan's poem frames from the outside as "sordid mortals". While this group has now lost its "powers", the loss suggests that they had "powers" to begin with and that they are based on this group's relationship with "nature". While Wordsworth's sonnet switches to a more personal "I" in the middle of line 9, distancing the speaker

---

19 Wordsworth, "The World Is Too Much with Us."

20 "Sordid" carries a complex set of meanings that I will not analyze in detail here, as they are also further complicated by being part of the paradoxical construction "a sordid boon" in Wordsworth's sonnet. My argument of similarity is based on a reading of "sordid" as a term that carries negative connotations, especially where it is used to describe human character or actions (see "Sordid, Adj. and n."). Its association with dirt also makes it a particularly useful term in the context of climate poetry, as it draws attention to anthropogenic impact on the environment as an act of agency – a deliberate "sullying" of the landscape.

21 Wordsworth, "The World Is Too Much with Us" (my italics).

from the "we"-group to present a personal solution for this sense of disconnect from "nature", the speaker's security in their own positionality as part of a "we" or as an autonomous self remains stable throughout. There may be different interpretations for this "we", ranging from all of humanity to a more specific group the poem tries to criticize for losing touch with "nature". There is, however, no doubt of the speaker's self-inclusion in this dominant group. By contrast, Logan's speaker is positioned outside of and in opposition to the "sordid mortals" who view "nature" with "cold indifference" and sacrifice "youth, ease, and health" to the pursuit of "Commerce".

While the speaker's positionality is not ostensibly gendered in either poem, I would like to suggest that a gendered reading of these differences suggests itself – not only based on a difference in opinion, but, more systematically in how gender (as a perceived and socially constructed category associated with specific traits and roles) emerges as a complication of positionality in Romantic writing concerned with the relationship of "the human" and "nature". This distinction is thrown into prominence in the work of women writers, who struggled with being recognized as writers as well as members of society.[22]

For the two poems I have discussed above, a gendered distinction suggests why the individual, emotional response of reestablishing a connection with "nature" rooted in a mythical, "Pagan" past that resolves the problem of disconnect in the sestet of Wordsworth's sonnet is not available to Logan's speaker, who carefully establishes a distance from both the "sordid mortals" engaged in commerce and an overly emotional immersion in "nature". As Anne Mellor suggests, "masculine Romanticism [...] glories in the triumph of the human mind over Nature" and a "brotherly" relationship, which ultimately "enable[s] the poet to *speak for* Nature."[23] I will further discuss this gesture, for instance in the context of Wordsworth's opposition to the Kendal and Windermere railway in my chapter on "Steam" below, distinguishing it from the more complex positionality of women writers around steam-powered mobility.

The focus of discussion in the following chapters will lie on gender as a dimension of positionality shaping responses to climate change and climate catastrophe in Romantic writing, as well as the critical discourses that have influenced and continue to frame current readings of the Romantics' and more contemporary engagements with anthropogenic climate change. These interwoven strands contribute to romanticized conceptualizations of "nature" and human control of the non-human

22 My focus here is not on the ways in which women writers interpreted this connection, but rather their experiences of seeing it destabilized in the context of climate change and climate catastrophe. For a closer look at women writers' poetics around "nature" see, for instance, Vargo, "Anna Barbauld and Natural Rights."

23 Mellor, "Thoughts on Romanticism and Gender," 344.

natural world that continue to resonate in contemporary discourses which I will address in my chapter on "Mushrooms" and general Conclusion below.

My discussion will draw on existing work on the ways in which climate change and climate catastrophe emerge in Romantic literature. In the last few years, there have been a number of critical studies discussing the Romantics' engagement with climate change, most prominently work by Seth Reno, David Higgins, Jeremy Davies, as well as Anne Collett and Olivia Murphy.[24] I will draw on these studies in tracing some of the most impactful phenomena, such as the climate crisis following the eruption of Mount Tambora in 1815. While these ecocritical and climate-oriented readings have slowly begun to include women writers of the Romantic period – most prominently Mary Shelley and Jane Austen – gendered positionality in and around climate change scenarios is not their main focus. Nor does there seem to be a concerted interest in interrogating the works of non-canonized women writers on climate change, as feminist criticism of Romantic writing has largely been focused on bringing critical attention to the contributions of women writers to the literary world of the Romantic period. These attempts are frequently focused on highlighting the underlying socio-political constraints that led to the omission of women writers from the canon of literary history, or on addressing central questions of literary production such as genre, poetics and aesthetics. As Mellor asserts in 2012, looking back on the work that had been achieved since the publication of her seminal *Romanticism and Gender* in 1993, there is still "a paucity of first-rate scholarship on the works of the women writers of the Romantic era."[25] Ten years later, this situation has begun to change, thanks to a continuous widening of the Romantic canon,[26] as well as global research connections and digitalization efforts, but there is still a lot of critical ground to be covered.

Despite the efforts of feminist literary historians in the 1980s and 90s to make the texts of women writers available to a broader readership, women's writing of the Romantic period is still frequently underrepresented in anthologies of Romantic writing[27] and not generally the main focus of overviews and introductions to Romanticism which tend to focus on the "Big Six" with an occasional chapter on women's writing[28] – a situation which seems to reflect critical omissions more than the productivity or popularity of women writers in the Romantic period. The texts collected

24 see Reno, *Early Anthropocene Literature in Britain – 1750 1884*; see Higgins, *British Romanticism, Climate Change and the Anthropocene – Writing Tambora*; see Davies, *The Birth of the Anthropocene*; see Collett and Murphy, *Romantic Climates: Literature and Science in an Age of Catastrophe*.

25 Mellor, "Thoughts on Romanticism and Gender," 346.

26 see Haekl, *Handbook of British Romanticism*, 1.

27 Duncan Wu's otherwise extensive *Romanticism*, for instance, includes only two sonnets by Charlotte Smith, a writer whose work was highly influential on Wordsworth. (see Wu, *Romanticism*.)

28 Mellor, "Thoughts on Romanticism and Gender," 346.

in anthologies dedicated to the work of women writers have not necessarily seen a major uptake in critical discussion – let alone in critical discussions of Romanticism and climate change.[29] Mary Ann Browne's "A World Without Water" and Joanna Baillie's "Address to a Steam-Vessel", which I discuss in my chapters on "Apocalypse" and "Steam", for instance, have seen no discernible critical discussion to date – in spite of their blatant topicality for the discussion of climate catastrophe in a period that is now generally understood as the early anthropocene.[30]

While I had originally planned to include a broader range of texts by women writers of the Romantic period, my own abilities to travel for research purposes were, somewhat ironically, hampered by my positionality as a disabled academic during the ongoing Covid-19 pandemic. The texts I discuss were, consequently, not discovered through extensive archival work, as I had originally planned, but gathered from anthologies of women's writing in the Romantic period.[31] While I can make no claims to completeness or even provide a broad overview in this way, the textual examples I draw on already suggest that Romantic women writers' engagements with climate change and climate catastrophe are an invaluable resource in foregrounding different views and strategies of thinking, writing and worldmaking in the face of climate crisis. Shaped by underlying discussions on contested questions of "humanity", "nature", and the experiencing and writing subject, women's writing of the Romantic period, raises questions that continue to resonate in contemporary discourses around anthropogenic climate change and its necessary mitigation. Most prominently, as I will argue below, in terms of its focus on a subject position that complicates the idea of the poet's ability to "speak for nature".[32]

In four chapters on "Anthropocene", "Steam", "Apocalypse" and "Mushrooms", I offer contrasting readings in the light of these questions by focusing on poetic, narrative, essayistic, autobiographical, ethnographic and popular science texts that engage with specific aspects of climate change or climate catastrophe from different perspectives.[33]

In Chapter 1 – "Anthropocene" – I trace critical feedback loops and connections between Romanticism and the anthropocene, as well as critical approaches to the

29 see Wilson and Haefner, "Introduction."

30 I will further discuss conceptualizations of the anthropocene and its potential starting point(s) in my chapter on "Anthropocene" below.

31 Mary Ann Browne and Joanna Baillie's texts were first discovered in Feldman, *British Women Poets of the Romantic Era.*; see also Wu, *Romantic Women Poets*; see also Ashfield, *Romantic Women Poets – 1770–1838*.

32 Mellor, "Thoughts on Romanticism and Gender," 344.

33 My focus lies on the ways in which gender emerges in texts focusing on climate catastrophe from the transatlantic Anglo-American context, in full awareness of the fact that it would be equally interesting and fruitful to include other forms of marginalization, based, for instance, on racializations or geopolitical location in the discussion.

anthropocene that allow me to interrogate the dominant Romantic narrative of "man"'s relationship with "nature" and its critical pervasiveness, which has led to the continued exclusion of women writers from the canon of Romantic literature. My argument is based on the assumed distance of the poetic subject from the observed aesthetic object in the eighteenth-century aesthetics of the sublime, which has created a pervasive understanding of "man" and "nature" as two distinct and separate entities. As I argue, this underlying aesthetic separation of "man" and "nature" has far-reaching consequences for humanity's relationship with the natural world in the understanding of the Romantics as well as our own. I trace some of the strategies that emerge in women's writing around "nature" and the aesthetic concept of the sublime, to show how women writers pushed against these pervasive mechanisms of exclusion. In the last part of the chapter, I present exemplary readings of three texts by Romantic women writers – Elizabeth Kent, Mary Shelley and Letitia Elizabeth Landon (L.E.L.) – to highlight genre-specific strategies and ways of negotiating positionality in their understanding of "nature" and climate change.

In Chapter 2 – "Steam" – I argue that the introduction of steam-powered travel – predominantly the railway – highlighted processes of acceleration and "progress" which, in retrospect, emerge as drivers of climate crisis. I trace reactions to these processes in Romantic writing, and show how the engagement with questions of progress, access and temporality allowed Romantic writers to reflect their own positionality. I argue that William Wordsworth embodies a universalizing perspective in his protest against a proposed railway through the Lake District in 1844. Henry David Thoreau's engagement with the railway in *Walden* (1854) is based on a position of situated knowledge that allows him to interrogate fundamental questions of temporality and the socio-political impact of progress. The third textual example in this chapter, Joanna Baillie's "Address to a Steam-Vessel" (1823), which I read in opposition to Wordsworth, challenges the dominant Romantic experience of "nature" as solitary and elitist in introducing steam-powered travel as a means of more democratic access to the landscape. I trace the question of access to the discussion of women's travel writing and the engagement with different versions of "nature" in women's writing in the concluding part of this chapter.

In Chapter 3 – "Apocalypse" – I trace Romantic reactions to the changes in climate triggered by the 1815 eruption of Mount Tambora, arguing that the fact that the Romantics were unaware of the causes of the climate catastrophe they were experiencing led to literary engagements with secular apocalypse and the questioning of aesthetic positionalities from which such events could be described. I read Lord Byron's "Darkness" (1816) as a critical engagement with the aesthetics of the sublime, arguing that only the introduction of a dream framework allows him to maintain a poetic subject position in the face of apocalypse. I compare "Darkness" and Mary Ann Browne's poem "A World Without Water" (1832) as distinct Romantic engagements with secular apocalypse, arguing that, in presenting climate catastrophe from an embodied perspective and supported by a plethora of structural elements, Browne's poem emerges as an example of situated positionality. The last part of the chapter presents a reading of the multilayered narrative structure of Mary Shelley's postapocalyptic novel *The Last Man* (1826). I argue that Shelley's novel negotiates Romantic questions around "nature" and positionality in its narrative structure and meta-commentary.

In Chapter 4 – "Mushrooms" – I trace romanticizations of mushrooms from the Romantic period to the current interest in all things 'fungal' in popular culture. I argue that two distinct narratives around mushrooms emerge in Romantic literature – one that reads mushrooms as a type of plant, to which other things can be compared, and one that understands mushrooms as mysterious entities which defer meaning. I trace these mushroom discourses through a number of texts and theorizations, arguing that current mushroom discourses in popular culture tend to envision mushrooms within a framework of heroic savior narrative criticized by Anna Tsing and Ursula Le Guin, whose counter models – theoretical and fictional – I also trace in this final chapter.

In the conclusion I return to the problem of feedback loops between Romantic conceptualizations of nature, Romantic ecocriticism and current romanticized approaches to climate crisis and the anthropocene that position "man" outside of "nature". My discussion throughout is situated on two different temporal levels, interrogating, on the one hand, the ways in which climate catastrophe shaped the Romantics' understanding of "nature", humanity and gendered poetics and, on the other hand, how Romantic constructions and poetic framings of "nature" and the human subject position still resonate throughout current framings of and reactions to anthropogenic climate change and climate catastrophe. Rather than present a unified, authoritative narrative of Romanticism, climate change and gender, my project attempts to piece together multiple ways of looking at climate change and climate catastrophe through the eyes of the Romantics, to introduce situated positionalities as an alternative to authoritative ones, and to trace continuities and lingering echoes of Romantic (and romanticized) discourses in current discussions of climate crisis.

Chapter 1: "Anthropocene"

Introduction: "Romanthropocene"

> Aerocene, Agnotocene, Anglocene, Anthrobscene, Anthroposcene, Capitalocene, Chthulucene, Democene, Econocene, Eremocene, Eurocene, Growthocene, Gynocene, Homogenocene, Kapitalocene, Manthropocene, Metropocene, Misanthropocene, Naufragocene, Necrocene, Neganthropocene, Northropocene, Occidentalocene, Oliganthrocene, Paleoanthropocene, Phagocene, Phronocene, Plantationocene, Planthropocene, Plasticene, Polemocene, Pyrocene, Sociocene, Sustainocene, Symbiocene, Technocene, Thalassocene, Thanatocene, Thermocene, and White Supremacy Scene.[1]

In recent years, there have been a number of attempts at establishing a critical connection between Romantic poetics, criticism and theorizations of the contemporary epoch,[2] which is now commonly referred to as "the Anthropocene" – even though, as I will argue below, this is not an entirely uncontested term. The focus of these studies varies of course, but there is an underlying sense of a continuity of Romantic ideas driving contemporary critical discussion – to the extent that Thomas Ford has suggested the term "Romanthropocene", adding it to the very impressive list of "-cenes" that might replace or supplement the term "Anthropocene" quoted above, to argue for an even closer, discursive connection between Romanticism and the anthropocene.[3]

Ford suggests that, while the "Anthropocene concept enacts an almost unthinkable break with prior historical periods", it also "remobilizes an extensive array of Romantic linguistic and cognitive forms, and draws on multiple Romantic genealogies."[4] His detailed historical account of the different components of the term con-

1 Ford, "The Romanthropocene," 1.

2 see Clemens, *The Romanticism of Contemporary Theory*; see Lörke, *Liminal Semiotics: Boundary Phenomena in Romanticism*; see Davies, "Romantic Ecocriticism."

3 Based on arguments which I shall further explore in this chapter, I am going to spell the term "anthropocene" with a small "a" throughout, even if it is usually capitalized in other contexts.

4 Ford, "The Romanthropocene," 2.

cludes in a crucial parallel between Romantic textuality and the anthropocene as an anthropogenic act of inscription on a planetary scale:

> In Schlegel's words, the philosophy of Romantic poetry shows "how the human spirit impresses its law on all things, and how the world is its work of art" (Schlegel, 1971, 183). The Anthropocene means that the natural world has been materially transformed by collective human history such that all earthly things now potentially bear the imprint of the modern human spirit. Nature then appears today only through a dense mist of historical traces that surround it on all sides [...], an emblem of a lost material freedom that has been ambiguously but comprehensively textualised. In this sense, of nature as an elegiac medium of bottomless historical inscription, the Anthropocene converges with Romanticism to give the title of the failed, melancholy, sombre, inconceivable artwork that has been made of the planet through the ubiquitous impress of human action: the Romanthropocene.[5]

While I would agree with the parallels and connections suggested by the term, and the mindset it attempts to characterize, the act of naming repeats a gesture that is not only elegiac and oriented towards a Romantic past but also seems intent on repeating an anthropocentric act of mastery and characterization that is a prevalent gesture in Romantic criticism and at least implicated in anthropocene criticism *critical* of the theoretical framing inherent in the term "Anthropocene". This, of course, also underlines the direct connection and the elegiac stance encompassed in the portmanteau term "Romanthropocene", suggesting that this may be the entire point of the exercise of coining yet another term.

The critical approach to the term can, by comparison, be retraced in an argument Ford makes earlier in the same article, which complicates the conceptual parallel of "the human spirit" and the "anthropos". Ford unpacks the doubling of the term "human" based on the different development of its two linguistic roots – Latin "homo" and Greek "anthropos" in Western thought: the "anthropological relationship that was first established in Romanticism increasingly split the knowing human subject from the colonised cultures that were the objects of that subject's field work."[6] As Ford acknowledges, this has far-reaching consequences based on the conceptual split between the meaning-making subject (homo) and the object of study and categorization (anthropos) which invites a discursive externalization of anthropogenic problems: "The anthropos of the Anthropocene is [...] is at once too universal, obscuring questions of climate justice, and too human, encoding technocratically Promethean modes of ecological and social management."[7] This subject/object split

5 Ford, 11.

6 Ford, 5.

7 Ford, 5.

also promotes a diffusion of responsibility, away from the "us" towards an imagined group of "anthropoi", whose impact on the planet can be studied and described. This may not be a conscious distinction, but its perseverance hampers direct action.

In this chapter, I want to take Ford's suggestion that "[w]hat the Anthropocene means depends at least in part on its Romanticism"[8] at face value, adding an additional critical complication by suggesting that what "Romanticism" means is also, historically, determined by Romantic criticism and that there are multiple versions of Romanticism in these critical accounts which need to be taken into consideration. This is not only and not predominantly a literary-historical project, but one that is also concerned with the Romantic legacy and the legacy of Romantic criticism in anthropocene discourse.

As Davies argues, early ecocritical Romantic criticism (Bate's work in the 1990s, for instance), "saw environmentalism as essentially a moral enterprise. Literature was for them a means to foster more caring and humble attitudes towards nature on the part of a universalised or generalised humankind."[9] This "'idealist' version of Romantic ecocriticism"[10] led to a lack of critical distance from the subject matter – the work of a relatively small group of canonized Romantic writers – and a focus on connections that fit into the critics' understanding of ecology. As Davies points out: "[c]oncentrating on how their chosen writers foreshadowed modern environmentalism made it harder for them to see those writers in their own literary and intellectual contexts, let alone in their social and environmental contexts."[11] Moreover: "their implied version of modern environmentalism had its own limitations",[12] in that it employed the work of the Romantics in a romanticizing ecological agenda. As Davies argues,

> In the critics' own age of pollution and rampant industrialisation, the Romantic ascription of intrinsic value to placehood, dwelling, and organic interrelationship has much to teach us about how to live in harmony with the natural world. On those grounds, the ecocritics roundly rejected the historicist desire to maintain a sceptical distance from "Romantic ideology."[13]

In this, they ran counter to a different school of Romantic critics drawing on Jerome McGann's highly influential *The Romantic Ideology* (1985) which postulated the necessity of critical distance – especially in Romantic criticism which McGann diagnosed

8 Ford, 2.

9 Davies, "Romantic Ecocriticism," 5.

10 Davies, 5.

11 Davies, 5.

12 Davies, 5.

13 Davies, 3.

with a tendency to get emotionally enmeshed in its subject matter.[14] I do not wish to rehash this historical debate here – nor do I wish to question its usefulness. Instead, I draw on Davies' historical overview of Romantic ecocriticism to draw attention to the impact of meta-critical feedback loops[15] which reiterate specific arguments, leading to their continuous reinscription in different cultural contexts. This is not simply a problem of twentieth- and twenty-first century Romantic criticism, but one that has, via the "Romanthropocene" connection, seeped into more contemporary anthropocene discourses around "the human" and "nature", which continue to be amplified by Romantic and romanticizing feedback loops. Simon J. White argues, for instance, that "Bate's broader claim that Wordsworth's Romantic nature aesthetic inspired, and continues to inspire, mainstream environmentalism in Britain is undeniably correct."[16] This has far-reaching consequences. As White argues,

> Romantic nature aesthetics continues to influence the way many British people look at rural spaces around them, and how they think about questions of environmental preservation or restoration. For many people, expansive green spaces, particularly national parks like the Lake District, constitute nature at its best and most abundant, despite the 2016 *State of Nature* report finding that in terms of biodiversity and wildlife decline Britain is now "among the most nature-depleted countries in the world."[17]

White's focus is on human and non-human presences in these landscapes, but his argument also highlights a continuity from Romantic, to Romantic ecocritical, to environmentalist thought that is central to my point here. As I will argue in more detail in my chapter on "Mushrooms", a Romantic/romanticized understanding of ecology celebrating the Romantics' unique understanding of and ability to emotionally and aesthetically connect with "nature" may not only overshadow other perspectives, it may also further cement an understanding of Romantic genius that might actually be detrimental to highlighting the full complexity and multi-perspectivity of current ecological exigencies.

This critique of historical environmentalist discourses has long been a concern of feminist ecocriticism. Val Plumwood's seminal *Feminism and the Mastery of Nature*

14 see McGann, *The Romantic Ideology: A Critical Investigation*.

15 I borrow the concept of "feedback loops" from climate science, where it refers to climate processes that are reinforced by their own effects feeding back into the system. As I argue, dominant ideas of "nature" in canonized Romantic writing and in current conceptualizations of climate crisis in the anthropocene reinforce each other via the Romanthropocene connection. On feedback loops in climate science see Masson-Delmotte, Pirani, and Connors, "Climate Change 2021: The Physical Science Basis."

16 White, "Untidying the Landscape: Romantic Poetics, Class and Non-Human Nature," 231.

17 White, 230.

(1993) for instance, opens on the argument that "'nature' has been a very broad and shifting category and has encompassed many different sorts of colonization."[18] For Plumwood the "domination of nature" is deeply entangled with other forms of oppression, which ecological feminism needs to address before it can establish itself against the "heavily masculine presence which has inhabited most accounts of environmental philosophy, including those of many deep ecologists."[19] Plumwood's historical scope is much wider than the one I am concerned with here, reaching from antiquity to postmodernism, but her assessment of continuities in ecocritical thought is nevertheless important to my discussion. Plumwood's argument is based on the assertion that "western culture has treated the human/nature relation as a dualism and that this explains many of the problematic features of the west's treatment of nature which underlie the environmental crisis, especially the western construction of human identity as 'outside' nature."[20] In this construction, women are frequently positioned on the side of "nature", rather than "the human" – relegating them to the position of aesthetic object rather than critical subject. More recently, Stacy Alaimo has argued that "[e]nvironmental discourses are often drenched with the legacies of the term 'nature'. Since the concept of 'nature' has long been enlisted to support racism, sexism, colonialism, homophobia, and essentialism, it remains a rather volatile term, which feminists should approach with caution."[21]

Romantic writing on "nature", especially in its critical reception, has fed into this conceptualization, as I have argued above. While the canonized Romantics generally seek an emotional and aesthetic integration with "nature", they do so from a vantage point that is originally outside of it. As Mellor shows in *Romanticism and Gender* (drawing on Wordsworth's poetry), in "masculine Romantic constructions of the sublime, a male poet finally speaks of, for and in the place of a nature originally gendered as female."[22] This experience is often universalized as quintessentially Romantic by Romantic criticism, further cementing a popular understanding of the human subject's position in and relationship with "nature".

Arguments countering such attempts at universalizing a very specific ecological perspective come from postcolonial[23] and feminist criticism, reimagining and challenging the Enlightenment concept of "the human" and its relationship with the "Others" marginalized in its construction and reiteration – including the non-human natural world.[24] As Plumwood argues, "[o]nce nature is reconceived as capable

18 Plumwood, *Feminism and the Mastery of Nature*, 1.

19 Plumwood, 2.

20 Plumwood, 2.

21 Alaimo, *Exposed: Environmental Politics and Pleasures in Posthuman Times*, 11.

22 Mellor, *Romanticism and Gender*, 90.

23 See for instance Ve-Yin Tee's introduction on "Environmentalism, Class and Nature" (see Tee, *Romantic Environmental Sensibility*, 1ff.)

24 see Gabrielson, "The Nature of Gender."

of agency and intentionality, and human identity is reconceived in less polarized and disembodied ways, the great gulf which Cartesian thought established between the conscious, mindful human sphere and the mindless, clockwork natural one disappears."[25] This is a central move for environmentalist projects: "if we do not understand the development and the defects in the western story of reason and nature, we may remain trapped within it or settle for one of its new versions."[26] While both Mellor and Plumwood were developing these ideas in the 1990s, contemporary critical discussions of climate change, mass extinction and their potential mitigation, in the feminist posthumanities build on their groundwork. I also draw on the work of Anna Tsing and Donna Haraway in this context, specifically in their critical discussion of anthropocene discourses, arguing against progress-driven narratives that reach back to Romanticism or draw on romanticizations of the relationship between "man" and "nature".

In the first part of this chapter – "The Anthropocene and its Critics" – I will introduce different readings of the term anthropocene as well as a number of anti-anthropocene arguments, to provide a critical framing for the following discussion of the connections implied in the term "Romanthropocene" between Romanticism, Romantic criticism and anthropocene discourses.

The second part of this chapter – "Romantic Criticism in the ANTHROPOcene" – will trace the emergence of gendered differences in critical discussions of Romantic writing in the anthropocene, and the ways in which these tend to further cement discourses around "women's writing" [27] from an exclusive critical perspective that dismisses their position as a form of self-domestication. As I argue, these distinctions form the basis for an understanding of gendered aesthetics that has a continuing impact on conceptualizations of "man"'s relationship with "nature" in the anthropocene.

In the third part of this chapter – "Gendering 'Nature' and the Sublime in Romantic Criticism" – I will draw on the work of feminist critics to show that strategies employed by women writers of the Romantic period in their engagements with "nature" complicate common critical narratives about the autonomous Romantic self's relationship with "nature".

In the last part of this chapter – "Women Writing 'Nature' and Climate Change in the Early Anthropocene" – I will take these questions to the discussion of different

---

25 Plumwood, *Feminism and the Mastery of Nature*, 5.

26 Plumwood, 6.

27 My purpose here is not essentialist in describing differences between women writers and their male counterparts or in cementing conceptualizations of "male" and "female" writing, but rather in tracing gender as a category of distinction that emerges in different contexts of Romantic writing and Romantic criticism and shapes further discussions of such fundamental concepts as "the human" or "nature".

textual examples, to show how these texts complicate critical narratives about the relationship between "the human" and "nature" in the Romantic period and that their inclusion in critical discussions might, consequently, contribute to a necessary critical revaluation of the "Romanthropocene" connection in Romantic criticism. This meta-critical discussion will also form the basis for my arguments in the following chapters, which read gender as a complicating factor in early-anthropocene discourses around climate change and climate catastrophe (focused on "Steam" and "Apocalypse"), before I return to the influence of these discourses and their connection to current conceptualizations of climate crisis in the anthropocene in the last chapter (on "Mushrooms") and the Conclusion.

The Anthropocene and its Critics

> The stories we tell about the world matter, but it is not always easy to know where to start or what to call them. [...] Changing the name changes the story. As, of course, does deciding where to start it.[28]

While the jury (of stratigraphers) is still out on whether and in what ways "the Anthropocene" might be recognized as a geological epoch, the term itself has already been widely adopted and used in a variety of climate change discourses, usually signaling an understanding of humanity's impact on the planet as so significant that it has changed its surface, climate and ecosystems. The *OED*[29] traces the term itself to a short article published by stratigraphers Paul Crutzen and E.F. Stoermer in the year 2000[30] in which they present an argument for naming a new geological epoch:

> Considering these and many other major and still growing impacts of human activities on earth and atmosphere, and at all, including global, scales, it seems to us more than appropriate to emphasize the central role of mankind in geology and ecology by proposing to use the term "anthropocene" for the current geological epoch. [...] To assign a more specific date to the onset of the "anthropocene" seems somewhat arbitrary, but we propose the latter part of the 18th century, although we are aware that alternative proposals can be made (some may even want to include the entire holocene). However, we choose this date because, dur-

---

28 Bould, *The Anthropocene Unconscious – Climate Catastrophe and Culture*, 6–9.

29 "Anthropocene, n. and Adj."

30 Crutzen and Stoermer, "The 'Anthropocene.'"

> ing the past two centuries, the global effects of human activities have become clearly noticeable.[31]

In the context of geology, this is a historical project focusing on the recognition of anthropogenic change in strata and ice cores. Crutzen and Stoermer postulate the year 1784 as the starting point of this era, arguing that the impact of the Industrial Revolution became detectable (in the form of several greenhouse gases beginning to show up in data retrieved from glacial ice cores), suggesting that anthropogenic climate change began to leave its marks on a planetary scale around this time. The term "anthropocene" in this sense is, predominantly, a scientific one, as Davies points out, and one that has a specific meaning for stratigraphers, even if it is also used in other contexts.[32] It is, consequently, important to distinguish between the scientific discussion surrounding the scope and validity of the term to describe specific phenomena in different scientific contexts and the parallel discussion concerning the usefulness of the term at the intersection of the environmental humanities and popular discussions of climate change and climate catastrophe to critically frame and envision human impact on a planetary scale.

While a rigid insistence on disciplinary boundaries is neither practical nor useful in the current context of global climate catastrophe – nor in attempting to describe the entangled system of cultural and emerging scientific discourses in the Romantic period – the instrumentalization and misrepresentation of scientific discussions by climate change deniers has made it necessary to constantly reiterate that a theoretical discussion of the usefulness of specific terminology to describe certain phenomena must not be confused with doubt about their existence and validity. In the same vein, the necessary critique of the term's usage in various contexts should not be confused with a general doubt of the fact that an anthropogenic influence on the planet, its ecosystems and climate can be detected on a major scale.

Climate change denial is, of course, seldom presented in good faith[33] and tends to muddle the issue in an attempt to undermine the credibility of scientific discourses and present a romanticized narrative of "man"'s dominant role over "nature". To illustrate this with a prominent example: on 26 February 2015 Republican senator James Inhofe brought a snowball to the floor of the US Senate to disprove the existence of global warming by showing that it was cold outside.[34] Inhofe did not cite any literature or, indeed, any observation of weather phenomena beyond the current day, but relied instead on the out-of-place materiality of the snowball in the senate chamber to make this point. While Inhofe's disingenuous argument

---

31 Crutzen and Stoermer.

32 Davies, *The Birth of the Anthropocene*, 2–3.

33 see Boström and Davidson, *Environment and Society*, 2.

34 *MSNBC – Senator Throws Snowball in the Senate.*

showcases a common, and in his case obviously deliberate misconception about the relationship between climate and weather in the service of partisan, capitalist interests, the frequency of such arguments in the political discourses surrounding climate change does not only highlight the complexity of the underlying scientific phenomena. It also draws attention to a certain argumentative resistance to acknowledging the uncomfortable existence of anthropogenic climate change and the ways in which social and political discourses about climate change tend to be driven by underlying assumptions, belief systems and interests. Moreover, as Ford argues,

> [o]ne major aspect of the conceptual challenges posed by the Anthropocene to existing disciplinary schemas of understanding has been that it requires thinking across vastly discrepant chronological scales and measures of change. It twins histories of modern technologies, and perhaps of literatures too, with those of the physical shaping of the earth's continents and oceans, its atmospheres and its lifeforms.[35]

Romantic theories of "the human" and "nature" constitute one network of meanings in this context, but one that, as Ford also argues had a lasting impact on current conceptualizations of the anthropocene. My primary focus in this chapter is the historical dimension of these discursive connections from the Romantic period onwards, as a critical recognition of the anthropocene in the context of literary studies opens up new perspectives and contextualizations not just for Romantic literature, but also for its connections to and influences on current discourses around anthropogenic climate change and climate catastrophe. This historical perspective traces discourse around the anthropocene back to a number of revolutionary changes introduced at the end of the eighteenth- and beginning of the nineteenth century and the ways in which they emerge in a variety of texts at the time.

As Reno, for instance, argues in *Early Anthropocene Literature in Britain*, a historical perspective on the (assumed) starting point of the anthropocene allows us to explore common questions in a new light and address the Romantics' awareness of climate change:

> [M]y contention is that the idea of the Anthropocene is older than popular opinion holds, and my focus is on the early years of the Anthropocene in the eighteenth and nineteenth centuries. Many of the debates and controversies of today surrounding industrialization, pollution, environmentalism, and globalism were already raging then, and while the texts and contexts are different, the root of the problem is the same: how to preserve the planet for future generations when humanity has become a geophysical force of nature.[36]

---

35 Ford, "The Romanthropocene," 3.

36 Reno, *Early Anthropocene Literature in Britain – 1750–1884*, viii.

Reno's own discussion begins in 1750, which, as he argues, "coincides with the Industrial Revolution, Agricultural Revolution, and rise of capitalism in Britain, and is thus an apt marker for the start of the global fossil fuel era."[37] It ends in 1884, with an epilogue on Modernism in the Anthropocene, and this choice of date is no coincidence.[38] On 4 and 11 February 1884 John Ruskin gave two lectures at the London Institution, in which he spoke at length about a newly observable cloud phenomenon that he described as "plague clouds", which turned the sky gray and dismal "in London, and at Oxford, through meager March, through changelessly sullen April, through despondent May, and darkened June, morning after morning."[39] As Reno argues, Ruskin not only "identified a new kind of atmosphere shaped by industrial pollution, coal smoke, and erratic weather patterns", his lecture also described "a new epoch dominated by a capitalist fossil fuel economy."[40]

Ruskin's lectures, while dismissed by his contemporaries as "the ravings of a once magnificent mind in decline,"[41] were not the only example of early anthropocene writing in Britain. Nor do they mark the beginning of this new epoch, which, by this point, had already been going for almost a century. His lectures are, in fact, only one example of an attempt at systematic description of existing phenomena. As Reno demonstrates, the notion that humans might have an impact on the climate was not necessarily as far-fetched as one might assume:

> French naturalist Georges-Louis Leclerc, le comte de Buffon attributed climate change to human activity in Epochs of Nature (1778). British author Thomas Carlyle wrote of "the Mechanical Age" in 1829. Scottish geologist Charles Lyell suggested the term "Recent epoch" in 1830. Welsh geologist Thomas Jenkyn wrote of the "Anthropozoic" in 1854, calling his present day "the human epoch." American environmentalist George Perkins Marsh summarized existing research on "climatic change" in *Man and Nature; or, Physical Geography as Modified by Human Action* (1864), writing that the industrial and agricultural revolutions of the eighteenth and nineteenth centuries "have tended to produce great changes in the hygrometric, thermometric, electric, and chemical condition of the atmosphere." Less than ten years later in 1873, Italian geologist Antonio Stoppani called this the "anthropozoic era."[42]

---

37 Reno, 4.

38 While I will, in part, draw on Reno's discussion of the Romantics, I am more interested in tracing the impact of Romantic era conceptualizations of climate change and climate catastrophe for our contemporary understanding of and engagement with these questions – and, most prominently, in a gendered perspective on these discussions.

39 Ruskin, "The Storm-Cloud of the Nineteenth Century," 3.

40 Reno, *Early Anthropocene Literature in Britain – 1750–1884*, 1.

41 Reno, 1.

42 Reno, 6.

As Reno's examples show, the term "anthropocene" can function as a useful framing to subsume these varied discussions and early writings on climate under a common heading and terminology, even before "Ruskin identified a new epoch dominated by the 'plague clouds' and 'plague winds' of industrialization" in 1884.[43] Reno's use of the term is directed at establishing a sense of historical continuity among a group of writers describing similar phenomena with a wide set of terminologies (even if some of them also identify the "anthropos" as a source of the changes they describe). This grouping, however, also limits the scope of the term to early (proto-)scientific writing in the European context, excluding other perspectives on these phenomena.

While the term "anthropocene" functions as a useful historical connector of early anthropocene climate writing, its implications are not entirely uncontested. The critique of its assumption of a generalized "anthropos" lies at the heart of critical discussions in the (post-)humanities, which have begun to question the usefulness of the term in discourses surrounding the current climate crisis and necessary mitigation efforts. Criticism is mainly directed at imbalances inherent in the term "anthropocene" itself, as well as its current use – most prominently its claim to universality (encompassing humans as a species without differentiating historically or in terms of geolocation or environmental impact), as well as its implied suggestion that it is something inherent in the species that is making "us" destroy the planet.[44]

There are several counterarguments against the term based on an attempt to shift its perspective to a critically more useful angle. As Haraway argues, "if we could only have one word for these SF times, surely it must be the Capitalocene. Species Man did not shape the conditions for the Third Carbon Age or the Nuclear Age."[45] As Haraway proclaims:

> Coal and the steam engine did not determine the story, and besides the dates are all wrong, not because one has to go back to the last ice age, but because one has to at least include the great market and commodity reworldings of the long sixteenth and seventeenth centuries of the current era, even if we think (wrongly) that we can remain Euro-centered in thinking about "globalizing" transformations shaping the Capitalocene. One must surely tell of the networks of sugar, precious metals, plantations, indigenous genocides, and slavery, with their labor innovations and relocations and recompositions of critters and things sweeping up both human and nonhuman workers of all kinds. The infectious industrial revolution of England mattered hugely, but it is only one player in planet-transforming, historically situated, new enough, worlding relations. The relocation of peoples,

43 Reno, 6.

44 I will further discuss Andreas Malm's counterarguments against this last thesis in my chapter on "Steam".

45 Haraway, *Staying with the Trouble*, 47.

> plants, and animals; the leveling of vast forests; and the violent mining of metals preceded the steam engine; but that is not a warrant for wringing one's hands about the perfidy of the Anthropos, or of Species Man, or of Man the Hunter.[46]

While Haraway highlights the elements of capital and commerce as the driving forces of the progress-oriented histories behind the Anthropocene/Capitalocene (a connection which I will discuss in my chapter on "Steam"), her counterargument also relies on a wider view of humanity as only one closely networked species on the planet. Haraway argues that the term "Anthropocene" suggests a much more dominant role for humanity than is warranted: "[t]he Capitalocene was relationally made, and not by a secular godlike anthropos, a law of history, the machine itself, or a demon called Modernity."[47]

A similar argument has been made by Heather Davis and Zoe Todd, in an article on the necessity of decolonizing the scientific arguments around the anthropocene. As Davis and Todd point out,

> the Anthropocene betrays itself in its name: in its reassertion of universality, it implicitly aligns itself with the colonial era. [...] the Anthropocene, if explicitly linked to the beginnings of colonization, would at least assert it as a critical project that understands that the ecocidal logics that now govern our world are not inevitable or 'human nature', but are the result of a series of decisions that have their origins and reverberations in colonization.[48]

Instead of reading (and thereby making excuses for) the current epoch as the result of an overly greedy "human nature", Davis and Todd propose that

> to use a date that coincides with colonialism in the Americas allows us to understand the current state of ecological crisis as inherently invested in a specific ideology defined by proto-capitalist logics based on extraction and accumulation through dispossession – logics that continue to shape the world we live in and that have produced our current era.[49]

Haraway's, as well as Davis and Todd's counterarguments against an implication of a universal "anthropos" in the creation of the current ecological crisis shift the focus away from humanity as a species and towards a very specific Eurocentric worldview

46 Haraway, 48.

47 Haraway, 50. Pointing forward, Haraway also introduces the term "Chthulhucene", to signal entanglements between various forms of life on the planet (see Haraway, "Anthropocene, Capitalocene, Chthulhucene.").

48 Davis and Todd, "On the Importance of a Date, or Decolonizing the Anthropocene," 763.

49 Davis and Todd, 764.

that is colonialist, capitalist and later industrialist but also patriarchal and that continues to treat marginalized groups of people, non-human animals, other organisms and the planet as a whole as resources to be exploited in the service of narratives of progress and the accumulation of capital. This framing points back towards a historical epoch that encompasses British and colonialist North American cultural discourses around 1800 in their engagement with commerce, industrialization and its impact on cultural understandings of "nature" and what it meant to be "human". Moreover, these attempts at reframing the anthropocene also challenge the conceptual division of the human into "homo" and "anthropos" and shift the blame away from the frequently colonized object of study (the anthropos) and closer to the investigating subject (the homo).

Haraway's, as well as Davis and Todd's arguments emerge from feminist and decolonial discussions that draw attention to the importance of situating knowledge and terminology within epistemological networks as well as networks of power. In these interconnected critical contexts, the genealogy and current meaning of terminology matters, because it has a direct and powerful impact on whether it is taken seriously and by whom. Haraway's and Davis and Todd's counterarguments, thus, draw attention to who is being excluded from scientific as well as cultural debates about the anthropocene and, by extension, whose contribution to the discussion might be valued as part of a scientific or activist context.

In *The Mushroom at the End of the World* (2015), which I will discuss in more detail in my chapter on "Mushrooms" below, Anna Tsing relates these historical framings to current discourses around climate catastrophe: "the assumption that the trope of progress is sufficient to know the world, both in success and in failure" is a fundamental part of contemporary culture. "Progress is embedded [...] in widely accepted assumptions about what it means to be human."[50] In the current situation, however, "[n]either tales of progress nor of ruin tell us how to think about collaborative survival."[51] Tsing argues that the "modern human conceit" that she sees as an inherent part of the current anthropocentric perspective of the anthropocene stands in the way of "noticing" other perspectives and entanglements.[52] Instead it insists on establishing master narratives that point away from the current moment towards "progress". While Tsing does not follow Haraway's critique of the term "anthropocene" or distinguish between "homo" and "anthropos" (she uses the term "anthropo-" to signal the dominant Western subject position), her argument adds an important component to my discussion here. Tsing draws attention to the fact that, in the context of Romanticism and its cultural legacy, these questions are inextricably linked to conceptualizations of humanity, "nature" and the Romantic

50 Tsing, *The Mushroom at the End of the World*, 21.

51 Tsing, 19.

52 Tsing, 19–20.

narrative of poetic genius as constitutive of how western culture understands the relationship between them. Not only is the "anthropo-" in the term "anthropocene" in its current use to broad in its claim to universality (including all of mankind), but it is also associated with patriarchally connoted narratives postulating a universal goal of constant "progress" that Tsing identifies as the driving factor behind the perpetuation of anthropocene problems. The universal subject position of "man" who stands in opposition to "nature" is the central character of these progress narratives that Tsing traces backwards to Enlightenment humanism, arguing that, "[t]his 'anthropo-' blocks attention to patchy landscapes, multiple temporalities, and shifting assemblages of humans and nonhumans."[53] Tsing's "anthropo-" corresponds to both parts of the homo/anthropos divide discussed by Ford, in that they both overshadow the many human and non-human "Others" marginalized and objectified by both labels.

The narrative of "man" with its focus on a singular way of perceiving and poetically engaging with "nature" also emerges in Romantic writing and Romantic criticism, overshadowing other forms of poesis, of perceiving and meaning-making. It also postulates a stance of objectivity towards aesthetic pluralities, narrowing critical perceptions down to one poetic system – which then excludes multiple or marginalized perspectives by default.[54] As I will argue in the chapter on "Mushrooms" and my Conclusion, these strategies of exclusion are not only deeply rooted in patriarchal systems, their reiteration through cultural-critical feedback loops also contributes to their cultural prevalence. They resurface in current discourses around climate change and climate catastrophe such as the gender climate gap[55] and other forms of climate injustice based on often intersecting dimensions of exclusion (such as racialization, class, geolocation, age or disability). These strategies are based on the postulate of a generalized figure of "man" under which all other positionalities can be subsumed (in the context of arguments about humanity as a whole, which often also use a universal "we") and judged or even excluded as exceptions or irrelevant marginalized perspectives. As my focus lies on gender here, I will continue with a closer look at Romantic criticism's exclusionary strategies on the basis of perceived gender differences in the following, to highlight some of the ways in which critical foci and categorizations contribute to the favoring and reiteration of specific narratives around "man"'s relationship with "nature" and the marginalization and dismissal of others.

---

53 Tsing, 20.

54 While I will predominantly focus on the example of women's writing in the context of my argument, this also holds true for other and, of course, intersecting forms of marginalization.

55 see Klus and Schramkowski, "Gender Climate Gap." Sebastian Klus and Barbara Schramkowski use the term "gender climate gap" in analogy to the "gender pay gap" to describe gender-based inequalities in the context of climate crisis.

Rather than trace a direct line from the exclusion of women from the Romantic canon via Romantic criticism to current marginalizations in climate discourses, I would like to highlight a specific feedback loop centered around what is perceived as politically and poetically relevant Romantic writing on "nature" by Romantic critics and which strategies might be employed in providing these exclusive definitions with a hue of objectivity. In a second step, I would like to further interrogate the dimensions of this definition, in terms of its focus on specific political and aesthetic markers – most prominently perspectives on "nature" and the sublime – which frequently exclude women's writing on the basis of perceived gendered differences, before I turn to concrete examples of the ways in which women's writing complicates the critical narrative of the relationship between the autonomous poetic self and "nature" in the last part of this chapter.

Romantic Criticism in the ANTHROPOcene

In *The Cambridge Introduction to Romantic Poetry* (2012) Michael Ferber makes a case for "Objectivity",[56] and against the multitude of subjective interpretations that he sees as emerging in the wake of poststructuralism and deconstruction: "Feminism, race-based approaches, postcolonial studies, and 'queer theory' have promoted a group identity version of this subjectivism arguing that there can be little commonality in the interpretations of, say, a straight white male and a black lesbian."[57] Ferber summarily dismisses these recent critical approaches, alongside New Historicism, suggesting that an underlying assumption of objectively recognizable meaning within the text is necessary to come to an interpretation of it and that, while some historical context may be "enriching", he ridicules the idea that "a work is 'a product of its time,' as if time ran a factory that bypassed real human creators."[58] His approach, thus, positions itself as vehemently opposed to the ecocritical readings of, for instance, Bate,[59] but also to more recent new materialist criticism interested in tracing the echoes of cultural and scientific contexts in Romantic writing.[60]

Ferber's claim of "objectivity" – rife with indignation about potential differences in perspective – does, however, not address the underlying historical and epistemological dimensions of the concept – most prominently what Haraway describes as the "deadly fantasy that feminists and others have identified in some versions of objectivity, those in the service of hierarchical and positivist orderings of what can

56 Ferber, *The Cambridge Introduction to Romantic Poetry*, 13.

57 Ferber, 14.

58 Ferber, 15.

59 see, for instance Bate, "Romantic Ecology Revisited."

60 see Davies, "Romantic Ecocriticism" for an overview of Romantic ecocritical work.

count as knowledge."[61] While the recommendation that any interpretation must be "defensible, and not just [someone's] personal 'response'"[62] sounds like sensible academic practice, Ferber's insistence on objectivity as a value in itself disregards the ways in which epistemological and academic hierarchies have historically shaped and continue to impact what can be understood as true or 'defensible'. Such a position also dismisses feminist, postcolonial or queer theory as mere "identity politics", disregarding the work that these critics have done to highlight historical exclusions and marginalizations in various disciplines.

As Haraway argues in her 1988 article on "Situated Knowledges", in the "realm of epistemology [...] what can count as knowledge is policed by philosophers codifying cognitive canon law", while, on the other side of a tipped scale "[a]cademic and activist feminist inquiry has repeatedly tried to come to terms with the question of what *we* might mean by the curious and inescapable term 'objectivity'."[63] Haraway's argument revolves around the question of objectivity in the context of the sciences, but her article also raises a number of points that are equally applicable to the humanities and, specifically, to literary criticism – most prominently the question of writing as a (gendered, racialized or otherwise) "embodied other" who, as Haraway summarizes, is generally "not allowed *not* to have a body, a finite point of view, and so an inevitably disqualifying and polluting bias."[64] In epistemological systems based on the Cartesian mind/body split, any sense of embodied knowledge automatically precludes objectivity.

To understand one's own position as "objective" based on its conformity with the historically ingrained academic default of cis, straight, white, male and able-bodied disregards the underlying power structures of academia. By extension, to read marginalized positions as political or politicizing disregards the political dimension of one's own position – especially, but not exclusively, in system of knowledge production. I am, however, not merely criticizing Ferber's all too common academic stance in failing to acknowledge his own positionality as a white male US citizen, anti-war activist and acclaimed academic who wrote seven books on Romanticism; or his failure to understand that his own positionality underpins a stance of critical authority that the students, who, as he claims, "believe that all that matters is their personal response to a poem", can not claim for themselves.[65] What concerns me

61 Haraway, "Situated Knowledges: The Science Question in Feminism and the Privilege of Partial Perspective," 580.

62 Ferber, *The Cambridge Introduction to Romantic Poetry*, 14.

63 Haraway, "Situated Knowledges: The Science Question in Feminism and the Privilege of Partial Perspective," 575.

64 Haraway, 575.

65 Ferber, *The Cambridge Introduction to Romantic Poetry*, 13.

about this stance, is the underlying problem of postulating objectivity without defining what that entails in the context of literary readings. To make such an entirely uncontextualized claim in an introductory text directed at students and published in 2012 signals a return to a postulate of objectivity that feminist critics – among others – have challenged for decades.[66]

In doing so, Ferber also reiterates a romanticized concept of universality – rooted in hierarchical systems of knowledge production and an assumed understanding of the relationship between subject and object, that, upon close inspection, also reveals itself as an ideological construct. The epistemological problem of how to approach objectivity without reproducing exclusionary gestures, as Haraway describes it, is much more complex and also includes the necessary work of thinking about

> how to have *simultaneously* an account of radical historical contingency for all knowledge claims and knowing subjects, a critical practice for recognizing our own 'semiotic technologies' for making meanings, *and* a no-nonsense commitment to faithful accounts of a 'real' world, one that can be partially shared and that is friendly to earthwide projects of finite freedom, adequate material abundance, modest meaning in suffering, and limited happiness.[67]

Haraway's argument draws attention to how closely the question of objectivity is tied to an understanding of the relationship between the observing subject and the observed object, one that aligns itself with questions about the relationship between "man" and "nature" – or to put this in less fraught terms, the observing and writing human being and the observed non-human natural world – in the Romantic era. By not acknowledging this dimension of objectivity, Ferber's argument, thus, also reiterates and reinscribes a number of historical discourses that complicate the discussion of Romantic writing about "nature".

I am reading Ferber's claims as a particularly blatant example of how Romantic criticism has historically engaged with women's writing of the Romantic period, arguing that his position reiterates a critical stance that not only reinforces the idea of a clearly defined Romantic canon but also justifies the exclusion of women writers based on a constantly reiterated sense of "objectivity" that he fails to interrogate. This is, as I have argued above, not simply a matter of criticizing a narrow reading of Romanticism, as Ferber's position also presents itself as an example of the ways in which individual critical texts form part of critical feedback loops, reiterating Romantic views in an argumentative structure that seems to draw on a Romantic un-

66 see Haraway, "Situated Knowledges: The Science Question in Feminism and the Privilege of Partial Perspective."

67 Haraway, 579.

derstanding of the concepts of "humanity" and "nature". I draw on feminist criticism and Romantic criticism to highlight these mechanisms and challenge Ferber's claim to "objectivity".

That a more nuanced and less reductionist discussion of objectivity not only continues to be necessary for the sciences – Haraway's focus – but also for the humanities, becomes particularly obvious in Ferber's chapter on "The Poetess",[68] in which he cites a number of women writers of the Romantic period whose "modest and deferential claims" about their own talent and position in the world seems to have automatically relegated them to the sidelines of Romantic literary production in his eyes.[69] This argument does not merely ignore the diverse approaches and longstanding and successful publishing careers of women writers of the Romantic period, it also instrumentalizes women writers' own gestures of humility and approaches to writing in writing them out of the critical picture.

Ferber postulates that the "thought of home with its duties and blessings, the wish to shun the glare of public renown and flourish in the light only of husband and family – these usually win out over the ranks of the poets shining in the heaven of fame."[70] This overly romanticized account of their decision making processes does not merely conflate the (perceived) predominant subject matter of women's Romantic poetry[71] with an acknowledgement of their own unimportance – reiterating a strict gendered separation of the public and domestic spheres that is rooted in nineteenth-century gender roles. Ferber's overly pathetic view of an entirely self-inflicted domesticity also sounds almost ironic in light of the struggle for women's legal rights that dominated the intellectual discussion at the time. Aidan Day's summary of these developments draws on Miriam Brody and Mary Wollstonecraft to highlight the ways in which William Blackstone's *Commentaries on the English Constitution* (1758) curtailed women's property rights to the point that they could not legally own property and did not have any rights over their own children or, indeed, any control of their own lives.[72]

It was, moreover, the harsh critical response to Wollstonecraft's political work which led, as Claire Knowles argues, to a "broader trend in women's poetry away from an engagement with radical politics in the first decades of he nineteenth century" largely as a result of a "generational change" as well as a "growing climate of fear – heralded by Powhele's [satirical poem] 'The Unsex'd Females', first published

68 Ferber, *The Cambridge Introduction to Romantic Poetry*, 36.

69 Ferber, 36.

70 Ferber, 37.

71 I will contest this assumption further below.

72 see Day, *Romanticism*, 20–21.

in 1798."[73] Political poetry, of which Anna Laetitia Barbauld remained one of the few representatives into the first decades of the nineteenth century, invariably had to engage with the impact of the French Revolution and the Napoleonic wars and the "two most famous female poets of the 1820s and 30s, Felicia Hemans and Letitia Landon, typically avoided overt critique of government in their poetry", a move which, as Knowles argues, "can also be seen, at least in part, as a survival strategy."[74] As Knowles argues with Judith Pascoe, the impact of and response to "Powhele's poem certainly 'helps modern critics demonstrate how impossible it was for late eighteenth-century women writers to be received in gender neutral terms.'"[75]

Ferber's "poetess" chapter seems to fall into the historical trap of accepting a cultural narrative that was anything but self-inflicted. His position further cements an understanding of who could be seen as a "poet" and who should be excluded from this rank, as it concludes with an observation that propriety would have clashed with genius in female poets, suggesting that "one can almost sympathize with this retreat into piety and propriety. But to repress the poetic genius is to repress Romanticism, and it is fair to ask whether the 'poetesses' who felt they should do so belong in a book about Romantic poets."[76] This somewhat reductive view, which denies women writers not only the title of "poet" but also that of canonized "Romantic" is underlined by Ferber's dismissal of all but four women poets, which he includes under the label of "canon" – Felicia Hemans, Charlotte Smith, Letitia Elizabeth Landon (writing as L.E.L. but listed here as "Letitia Landon") and Mary Robinson.[77]

The fact that it is possible to blame "ancient patriarchal laws and customs"[78] for this critical dismissal of an entire group of authors based on their gender presentation and then reinscribe it in the form of literary criticism, highlights the importance of further discussion of the exclusivity of Romantic ideas of poetics and authorial genius and the ways in which they have shaped and continue to shape contemporary modes of understanding and framing related ideas, such as innovation and ingenuity. Moreover, Ferber's exclusion of women writers based on his understanding of "domesticity" also draws on a binary (frequently gendered) model of inside/outside or home/world in which "domesticity" precludes any engagement with "nature" as the central topic of Romanticism. I will address this problem in the chapter on "Steam" below, where I also discuss questions of access and the ways in which women writers' engagements with "nature" were often not framed as such.

---

73 Knowles, "Female Romantic Poetry, 1798–1819: The Climate of Fear and the Loss of a Radical Generation," 306–7.

74 Knowles, 317.

75 Knowles, 306.

76 Ferber, *The Cambridge Introduction to Romantic Poetry*, 40.

77 Ferber, 6.

78 Ferber, 40.

Gender distinctions are central to these concerns, as for Ferber's argument about the Romantics and Romantic criticism, "[p]oets are also men", the "creator[s] of imaginative worlds." His observation that this is a "Romantic glorification of poets"[79] fails to acknowledge the reiteration of gender stereotypes and the conflation of ethics and poetics in his own critical approach. This argumentative move relies on what Manu Samriti Chander has aptly described as "numerous endlessly quoted, self-consecrating Romantic formulations", frequently reiterated throughout Romantic criticism,[80] and which I have attempted to describe as a feedback loop model here. Chander's own project in *Brown Romantics* (2017) focuses on colonial hierarchies, reiterating that "an asymmetrical balance of power and periphery" led to a marginalization of colonial poets in a manner that is both indicative of racial hierarchies and symptomatic of hierarchical exclusion. In labeling colonial poets as "brown", Chander argues,

> I want to ironize what might, at face value, be taken as a rather crude descriptor of racial difference, and, by thus calling attention to racial identity challenge that basis for considering their poetry as a simple expression of it. To put it in another way, "Brown Romantics" are not marginalized because they are brown; on the contrary, they are "brown" because they are marginalized. Their capacity to participate as poets in the cultural field is restricted by their relationship to the colonizer prior to their participation in it, which in turn exposes the racial dimension of the "world republic of letters."[81]

Chander's argument highlights the mechanisms of exclusion at work in the Romantic "world of letters", drawing attention to the circularity of these constructions. My intent in citing his work is not to conflate racist and patriarchal structures and claim that the same processes are at work in the exclusion of different but frequently intersecting marginalized groups, but rather to draw on his argument to further highlight the intricacies of critical feedback loops and to call attention to the historical dimension of these questions, which continues to impact who is in- or excluded in processes of canonization and periodization.[82]

Recent Romantic criticism, such as Chander's book, has begun to carefully disentangle the work of marginalized writers of the Romantic period from the exclusive verdicts of canonizing criticism. Norbert Lennartz, for instance, cites a staggering

79 Ferber, 16–17.

80 Chander, *Brown Romanticism – Poetry and Nationalism in the Global Nineteenth Century*, 2.

81 Chander, 3.

82 Such processes of marginalization, work in two directions, as I have argued elsewhere, excluding not only the marginalized Other from the hegemonic discourse, but also shaping and upholding the dominant subject position of the Anglo-European colonizer in the same process (see Heise-von der Lippe, *Monstrous Textualities*.).

number of 4000 writers producing poetry in Britain between roughly 1770 to 1835.[83] His *Lost Romantics* (2020) is "interested in [the] nondescript realm of hellish lostness", to show that "the two realms of canonicity and non-canonicity [...] are more closely related than hitherto assumed."[84] To frame the six "famous white men"[85] as "survivors",[86] as Lennartz does in his introduction, shifts the focus away from ideas of genius towards critical mechanisms of inclusion and exclusion that led to the historical canonization of this small group of writers.[87] Ferber, by comparison does not only reiterate a very specific, narrow canon, arguing that "[t]oday most anthologies and university courses dwell largely on the 'Big Six' male Romantic poets",[88] he also frames efforts of questioning or broadening this canon as a form of ongoing "assault" since its establishment "about 1950."[89] His assertion of a specific canon is based on an assumption of unchangeable, objective poetic merit that oversimplifies the complex poetic discussions the Romantics – and not just the canonized Romantics – were engaged in. That Ferber does so in 2012, reiterates the necessity of drawing attention to these constant processes of reinscription, as they continue to present a specific model of the Romantic worldview that resonates with a current understanding of "man"'s relationship with the non-human natural world in the context of climate crisis.

To underline his point, Ferber cites Percy Shelley's claim (in his essay "Defence of Poetry", written in 1821) that "poets are the unacknowledged legislators of the world"[90] – reading Shelley's argument as a claim of authority: "poets are the vehicles of the spirit of the age and the age to come."[91] Any claim to objectivity over what this means preempting a discussion of how this frequently quoted line came to mean this and for whom, constitutes a very narrow understanding of how meaning is created and reiterated in and through power structures and the positionality of the speaker. This is not a gesture unique to Ferber's text, as he, in fact acknowledges, but one which is often employed in the service of reinscribing a specific notion of the self-

83 see Lennartz, "Probing the Realms of Lostness, Non-Canonicity and Oblivion: An Introduction," 4.

84 Lennartz, 6–7.

85 Lennartz, 3.

86 Lennartz, 4.

87 Digital Humanities projects, such as Franco Moretti's concept of "distance readings" have also contributed to the idea of broadening the canon beyond the texts that are usually read as representative of Romantic literature (see Moretti, *Distant Reading*.).

88 Ferber, *The Cambridge Introduction to Romantic Poetry*, 6.

89 Ferber, 6.

90 Shelley, "A Defence of Poetry," 956.

91 Ferber, *The Cambridge Introduction to Romantic Poetry*, 16. Chander instead takes the legal connotations of the term "legislator" as a gateway into a discussion of the entanglements of colonial politics and poetics (see Chander, *Brown Romanticism – Poetry and Nationalism in the Global Nineteenth Century*.)

justification of the Romantic canon that I am criticizing here. Without the context of Shelley's text, the position of the "legislator of the world" might easily be misread as a claim to absolute authority, which is, indeed, not actually the argument Shelley makes.

Shelley's "Defence of Poetry" does not only revolve around the central question of how meaning is created through language, but also around the role of poetic language in the creation of the future, that is with "legislators and prophets"[92] and their status in comparison with those who "measure the circumference and sound the depths of human nature with a comprehensive and all-penetrating spirit, and they are themselves perhaps the most sincerely astonished at its manifestations."[93] The focus of Shelley's "Defence of Poetry" lies, as Mark Sandy points out, on the "rebuttal of Peacock's arguments about the redundancy of poetry."[94] The essay is, thus, not only an attempt at describing the function of poetry in worldmaking and the role of the poet as a moral guide of society, but also an entry into an ongoing discussion of these questions. Shelley reads poetic language as a form of epistemology and ethics, whose roots are neo-platonic in their conflation of poetic truth with beauty and goodness:

> The functions of the poetical faculty are twofold: by one it creates new materials for knowledge and power and pleasure; by the other it engenders in the mind a desire to reproduce and arrange them according to a certain rhythm and order which may be called the beautiful and the good.[95]

Worldmaking, in this sense, is a form of basic human artistic practice that Shelley compares to music and dance. Poets are "the unacknowledged legislators of the world" in that they are, fundamentally, concerned with creating it. The expression of this practice is mimetic and achieved through approximation not mastery. The truths thus created are, to an extent, universal, but may be expressed differently by different poets. While

> all men observe a similar, they observe not the same order [...] in the series of their imitations of natural objects. For there is a certain order or rhythm belonging to each of these classes of mimetic representation [...]. The sense of an approximation to this order has been called taste by modern writers.[96]

92 Shelley, "A Defence of Poetry," 946.

93 Shelley, 956.

94 Sandy, "Defence of Poetry."

95 Shelley, "A Defence of Poetry," 952.

96 Shelley, 945.

The essay does not seem to judge these distinctions of taste, but it stipulates that only those in whom the "faculty of approximation to the beautiful" exists "in excess are the poets, in the most universal sense of the word – and the pleasure resulting from the manner in which they express the influence of society or nature upon their minds, communicates itself to others."[97] While Shelley's remarks are contextual, they are nevertheless useful in the way they reflect the Romantics' stance on poetic genius: Shelley highlights the faculties of the poet in recognizing and expressing beauty (and by extension truth) but does not further specify who can inhabit the position of "the poet".

Romantic criticism, based on eighteenth-century aesthetic theory and nineteenth-century gender roles, adds a dimension of gender to these observations that is not foregrounded by Shelley himself – that is the distinction between the sublime (associated with the spiritual growth of the poet-genius – usually assumed to be male) and the beautiful object of his observation and writing (usually assumed to be female). I will address this distinction and its ramifications for the construction of Romantic subject positions in the face of climate catastrophe in the chapter on "Apocalypse" below. My argument here revolves around the construction of the figure of the poet in Romantic writing and Romantic criticism – and specifically the emergence of gender as a category of exclusion from the experience of "nature" in the critical reconstruction of Romantic poetics. Ferber's instrumentalization of Shelley's poetics in an attempt at postulating the male "poet" and excluding the female "poetess" reiterates an understanding of gendered poetics and politics that has not gone entirely uncontested in feminist criticism but, nevertheless, continues to reemerge in Romantic criticism in the twenty-first century. As Mellor argues in *Romanticism and Gender*,

> [m]asculine English Romanticism has long been associated with a love of nature, or more precisely, with the epistemological relationship of the perceiving mind to the object of perception. When the fully conscious poetic mind grasps nature that is entirely unmediated by language – or wholly constructed by its own linguistic tropes – it experiences what the Romantic writers called "the sublime". This concept of the sublime [...] is distinctly, if unwittingly gendered. The sublime is associated with an experience of masculine empowerment; its contrasting term, the beautiful, is associated with an experience of feminine nurturance, love and sensuous relaxation. This gender differentiation was implicit in the most famous treatise on the sublime published in the eighteenth century, Edmund Burke's *Philosophical Enquiry* [...].[98]

97 Shelley, 945.

98 Mellor, *Romanticism and Gender*, 85.

These gendered distinctions travel from Burke's aesthetics to Romantic writing only to be picked up and reiterated by Romantic criticism, where they are uncritically used to exclude writers and writing that do not conform with them or, at least, to postulate their work as a curious exception to an aesthetic rule. I will return to feminist criticism of this distinction in the next part of this chapter.

Based on its roots in the Burkean aesthetics of the sublime, mainstream Romantic criticism locates the possibility of experiencing the sublime in the extraordinary capacity of perception and expression that is constructed as poetic genius. This figure of poetic genius is invariably read as male, based on an understanding of the process of sublimation which is imbued with an understanding of male agency in Burke's aesthetics. While, as I have shown above, poetic genius emerges in Shelley's "Defence of Poetry", the stipulation that poetic genius must be male-connoted is a critical addition that has attached itself in the context of canon formation, where it reiterates the exclusion of perceived "lesser" writers and presents the genius poet as an autonomous figure, free of any attachments, except, perhaps to other figures of poetic genius.

As Higgins argues, "[g]enius was one of the concepts fuelling the cultural revolution that we now call Romanticism."[99] By the end of the eighteenth century, the modern usage of the term, signifying, according to the *OED*, "native intellectual power of an exalted type... instinctive and extraordinary capacity for imaginative creation, original thought, invention, or discovery",[100] was well established and "genius had become a key term across Europe in aesthetics, literary criticism, and accounts of subjectivity."[101] Higgins describes a shift towards more personal and private discussions of genius that occurred between the late eighteenth and early nineteenth century, arguing that a "great deal of writing on genius in the 1820s and 1830s was biographical, for the Romantic emphasis on the individual consciousness behind artistic creation – and on the exceptional nature of genius – contributed to an increasing fascination with the personalities and private lives of creative artists."[102] These accounts frequently form the roots of Romantic canon formation, reiterating views on specific, outstanding writers. While genius is not exclusively male coded in his view, Higgins cites Marlon B. Ross' argument that "the reliance on 'masculinist metaphors of power' in the writings of the Romantic poets and their supporters was a defensive response to the perceived challenges to the poetic vocation, perhaps most crucially the increasing prominence of women writers and readers."[103] Interestingly, contemporary

99 Higgins, *Romantic Genius and the Literary Magazine: Biography, Celebrity, Politics*, 2.

100 "Genius, n. and Adj."

101 Higgins, *Romantic Genius and the Literary Magazine: Biography, Celebrity, Politics*, 2–3.

102 Higgins, 3.

103 Higgins, 6.

> [c]ritics sometimes applied the term 'genius' to successful women writers like Joanna Baillie, Maria Edgeworth, Madame de Staël and Letitia Landon, but it generally seems to have been assumed by male writers that female genius, if it existed, was both less powerful and less extensive than male genius and that women wrote best when they confined themselves to sentimental subjects.[104]

While this assertion and its passive voice construction seem to resonate with Ferber's dismissal of the "Poetess", Higgins is more concerned with presenting a differentiated view of Romantic genius than with objectively reasserting Romantic distinctions in an attempt at reiterating a specific canon.

I am presenting these distinctions here, to draw attention to how Romantic poetics became instrumentalized in processes of canon formation, which reiterate specific views on the role of the poet and "his" relationship to the non-human natural world. In Romantic criticism, the figure of poetic genius is imbued with a particularly keen understanding of the natural world and a unique way of expressing it. This assumption translates itself into an almost prescriptive model in Romantic criticism, which, through reiteration, cements the view that this is, indeed, the only way in which "man"'s relationship to "nature" can and should be understood.

In his discussion of Romanticism as a contested critical term, Day draws on René Wellek (1949) to argue that Romantic criticism has historically been focused on establishing an argument for coherence concerning the idea that the "same conception of nature and its relation to man" can be found in the canonized Romantics.[105] One of the strategies he identifies is the foregrounding of the philosophical dimension of poetics. Carol Shiner Wilson and Joel Hafner describe that M.H. Abrams in *The Mirror and the Lamp* (1953) and *Natural Supernaturalism* (1971) posited "that true poets are philosophical. That philosophical role is, implicitly, public and male."[106] Wilson and Haefner do not only highlight the role of influential critics like Abrams in constructing a specific image of Romanticism, but they also draw attention to the ways in which critical narratives privilege certain features, such as the role of the Romantic poet as public philosophical genius whose insights shaped the Romantic worldview and its critical afterlife.

These narratives of the poet's unique subject position do not only tend to conflate the work of fairly distinct writers into a unified Romantic ideal, as Day argues, but they also tend to either subsume or further exclude marginalized writers on the basis of a coherent ideal of the poetic subject position that they don't seem to fulfil. As Day,

---

104 Higgins, 6.

105 Wellek quoted in Day, *Romanticism*, 4.

106 Wilson and Haefner, "Introduction," 4.

argues, this isolation of the genius poet from "his" historical context is also a problem of critical accuracy in describing Romantic innovations and continuities:

> It is this kind of focus on, and celebration of, subjectivity that is sometimes seen as the distinctive Romantic innovation. But while the 'extreme assertion of the self' (Drabble 1985: 842) and the use of nature 'as a stimulus for the poet to engage in the most characteristic human activity, that of thinking' (Abrams 1993: 128) are crucial features of Romantic writing, they didn't suddenly appear with the publication of *Lyrical Ballads* in 1798.[107]

Day's argument is based on the historically unacknowledged influence of Charlotte Smith's work on Wordsworth and Coleridge. The continuity of poetic traditions is, however, not the only casualty of critical arguments focused on self-assertion and the poet's unique mental capacity to process "nature". As Daisy Hay shows for the Byron-Shelley circle, community, and specifically literary cooperation and mutual influence, are often also lost in critical or biographic narratives that focus on poetic genius – leading to an imbalanced view of the individual as creative genius in romanticized narratives.[108]

I have presented these critical discussions on the emergence of the concept of poetic genius in Romantic writing in some detail here, to draw attention to the ways in which Romantic criticism, in its attempts at canon formation and at formulating a singular narrative about the Romantics' understanding of the central figure of poetic genius, has created a narrow and exclusive understanding of a diverse and complex set of poetic concepts in the works of the Romantics themselves. I have also attempted to show that recent overviews of Romantic writing like Ferber's introduction reiterate this narrow understanding, further cementing a specific view of the Romantics and their relationship to "nature" which continues to resonate with more recent conceptualization of "nature" in the anthropocene. As I have argued, feminist epistemology criticizes the dominant narrative of "man"'s relationship to "nature", arguing for a more complex, multifaceted understanding than the unified view that emerges from the model of Romanticism constructed by Romantic criticism. In the next part of this chapter, I will present a number of feminist critical reevaluations of Romantic writing as a basis for the readings of individual texts in the last part of this chapter.

107 Day, *Romanticism*, 42.

108 Hay, *Young Romantics*, 300.

Gendering "Nature" and the Sublime in Romantic Criticism

In the preceding part of this chapter, I have framed processes of critical exclusion that marginalize women writers of the Romantic period from a critically constructed subject position which reiterates the capacity of the male poet to relate to and speak for an externalized conceptualization of "nature". While the central figure of poet genius is imagined to be able to connect to "nature" in the process of perceiving and writing, this is a process of mastery which bridges an originally imagined separation. This understanding of the relationship between "man" and "nature" continues to resonate in unforeseen ways throughout current conceptualizations of ecology, because its imagined subject position that resides outside of "nature" reiterates binary models of the separation of "nature" and culture, humanity and ecology. As I have argued above, feminist epistemologies criticize this separation, arguing for more complex ways of situating knowledges in the experience of the non-human natural world. In this part of the chapter, I will present counterarguments against the critical narratives that I have outlined above, most prominently with regard to women writer's engagements with the concept of "nature" and the ways in which they complicate the critical narrative of mainstream Romanticism.

In *Romanticism and Gender* Mellor juxtaposes a "masculine Romanticism" and a "feminine Romanticism", which is not tied to the perceived gender of the poet but based on an understanding of the ways in which they engage with discourses around the sublime. As I have argued above, Burke's conceptualization of the sublime, which the Romantics largely drew on, is inherently male-connoted and tied to a process of sublimation that reiterates the agency of the autonomous self. In the Burkean sublime, there is no female subject position, as femininity is conflated with the beautiful as an object of contemplation, not agency. Mellor argues that women writers of the Romantic period challenge this gendered distinction in the ways in which they engage with the sublime. While in the masculine Romanticism of canonical Romantic poetry "nature" is identified "as the external objective world which the self-conscious subject must penetrate, possess and interpret",[109] the women writers Mellor discusses present the sublime as "an ecstatic experience of co-participation in a nature they explicitly gender as female [...]. Hence the tradition of the sublime for this tradition of women writers is rarely solitary."[110] This is not a straightforward dichotomy, however, as Mellor's own discussion also shows counterexamples and acknowledges the complex challenges women writers of the Romantic period faced in engaging with an already gendered discussion of sublime aesthetics. Letitia Elizabeth Landon's poem "Love's Last Lesson", which, Mellor also discusses, presents such a counterexample of sublime aesthetics, as it

---

109 Mellor, *Romanticism and Gender*, 21.

110 Mellor, 97.

> appropriates the traditional Western cultural identification of Nature as female, of the female as body rather than mind, as nature rather than culture, only to turn it in upon itself. In *Love's Last Lesson*, love's ending is troped as the "history of some fair southern clime," as geology and botany and biology. But instead of writing the female as Mother Earth, all-procreative, abundant and fertile, Landon writes the love of the "natural" female and female nature as a volcano erupting onto a paradisiacal landscape, laying "all waste before it," bringing not fertility but "pestilence" and "ashy waters" and "desert barrenness".[111]

Mellor's discussion of gender, while focused on establishing a distinction between masculine and feminine styles of approaching "nature" also draws attention to the complexity of aesthetic approaches in women's writing of the Romantic period which were, conceptually as well as socially, excluded from the role of poetic genius. Before the imagined descent into volcanic chaos, Landon's speaker "[flings] aside the scroll, as it had part / In her great misery. Why should she write? / What could she write?"[112] This act of defiance and despair – directed at the lover who has scorned her – also functions as a form of meta-poetic commentary, as the overt speaker, bemoaning her loss, switches to a more detached third-person description of her poet alter ego's possibilities of engaging with this loss in writing which "might not frame her utterance" but can imagine "lay[ing] all waste before it."[113] To dismiss Landon's meta-poetic engagement with the ways in which emotions can emerge in poetry and to what extent poetry is able to encompass them aesthetically as a deliberate act of shrinking or self-domestication (as Ferber does), requires a conscious act of suppression. This specific critical move, however, also draws attention to the perseverance of discourses of Romantic poetics in Romantic criticism – specifically the impact of the lone figure of poetic genius on literary histories and processes of canon formation.

As Stephen C. Berendt argues the "literary historical account" was not only dominated by "more patrician tastes" that excluded "laboring-class" poets, "the greatest exclusion from the canon [...] involved women, whose names and poetry were largely absent from discussions of Romantic poetry for much of the twentieth century."[114] Berendt's argument draws attention to the ways in which such exclusions shape and are shaped by processes of literary canon formation. He, however, also highlights the pervasiveness of such frequently institutionalized misrepresentations,[115] which he

---

111 Mellor, 118.

112 Landon, "Love's Last Lesson."

113 Landon.

114 Berendt, "Poetry," 1.

115 Berendt, 2.

frames as "revisionist literary history", actively engaged in obliterating the contributions of women to Romantic writing.[116]

Berendt's argument draws on Stuart Curran's 1988 essay "The I Altered" – one of the first attempts at resurrecting the work of women writers of the Romantic period from this deliberately constructed critical exile. While Curran's essay was instrumental in the "rediscovery" of a number of popular women writers of the Romantic period, it also somewhat paradoxically focuses both on the celebrated poetic genius of women writers *and* their unique (gender specific) perception – that is, their "fine eyes who are occupied continually in discerning minute objects or assembling a world out of its disjointed particulars."[117] In this, Curran's argument seems to foreground their aesthetics rather than their abilities, also suggesting a direct route from noticing to writing that seems to – purposefully or inadvertently – omit the detour via the mind and its ability to engage with and process the discerned objects. This early attempt at reframing the critical narrative around women's writing of the Romantic period, thus, reiterates binary stereotypes drawing on women's associations with the beautiful, instead of the sublime and essentially locating their poetic capacities in the body instead of the mind.

Male writers were, on the other hand, often celebrated precisely for their capacity of reflection. Day, for instance, quotes Abrams as saying:

> It is a mistake, however, to describe the romantic poets as simply 'nature poets'. While many major poems by Wordsworth and Coleridge – and to a great extent by Shelley and Keats – set out from or return to an aspect or change of aspect in the landscape, the outer scene is not presented for its own sake but only as a stimulus for the poet to engage in the most characteristic human activity, that of thinking.[118]

Where women, thus, might be physically more able or just better trained to "notice" things (as Curran suggests), the element of thought is omitted from his description. This gender-based juxtaposition of physical and mental abilities, thus, highlights far more than the injustice of their exclusion from the critical canon. To ascribe an enhanced capacity of "noticing" to women simply because they are women would be an oversimplification of much more complex processes of identity formation and the way socialization shapes perception and its expression. Such an attribution is also a kind of romanticization in itself (one that is frequently also made for other

116 Berendt, 2.

117 Curran, "Romantic Poetry: The I Altered," 189.

118 Abrams quoted in Day, *Romanticism*, 127–29.

marginalized groups upon their "rediscovery" by the mainstream[119]). Curran's argument is based in second-wave feminist essentialist thought patterns that locate gender distinctions in the body, but also reiterates some of the aesthetic framings inherent in Romanticism and Romantic criticism, most prominently the binary structure of sublime/beautiful. The shift to a social and performative understanding of the conception of gender in third-wave feminism contests these binary structures. Mellor who is, similar to Curran, also engaged in the central work of putting women's writing back on the critical agenda, highlights the difficulty of doing so based on an insistence on specific qualities of "women's writing" – let alone women's physical abilities. Instead of ascribing special faculties of noticing to women writers as Curran does, Mellor focuses on the ways in which women writers attempted to be taken seriously by focusing on "the workings of the rational mind, a mind relocated – in a gesture of revolutionary gender implications – in the female as well as the male body."[120] This focus on rationality emerges, in Mellor's view, in response to the concerns of

> their male peers with the capacities of the creative imagination, with the limitations of language, with the possibility of transcendence or 'unity of being,' with the development of an autonomous self, with political (as opposed to social) revolution, with the role of the creative writer as political leader or religious savior.[121]

The question of who was seen as a rational subject and whose subjective vision shaped the epistemological possibilities of understanding and poetically envisioning literary framings of the world is not a trivial one – especially not for a literary culture that, to a large extent, understood poets to be in Percy Shelley's sense "the unacknowledged legislators" of the – natural as well as political – world.[122] The critical gesture of acknowledging women's work, only to place it in a distinct category separate from the work of canonized male poets, however, tends to deepen the gulph of omission.

As Wilson and Haefner (writing in 1996) point out with Paul Cantor, the revaluation of various 'minor' writers of the Romantic period is a relatively recent project and one that relied on discrediting the "Romantic idea of the artistic genius".[123] Curran's article can be seen as part of this process, but his careful explanation of the specific gendered differences of women's writing also offers some insights into why they

119 Echoes of Rousseau's "noble savage" still prevail in environmentalist discourses surrounding Indigenous conservation and restoration practices, which are often misrepresented or romanticized.

120 Mellor, *Romanticism and Gender*, 2.

121 Mellor, 2.

122 Shelley, "A Defence of Poetry."

123 Cantor quoted in Wilson and Haefner, "Introduction," 4.

might have been excluded from critical narratives based on a framing that reiterated gender stereotypes about domesticity and the ways in which women engage with the outside world. Curran's approach of drawing attention to the unique poetic abilities of women writers of the Romantic period was helpful in rekindling critical attention for Romantic landscapes of publishing and reception that did not only include but also value the work of writers who had been excluded from the canon of literary history. His article, however, also served to further cement assumptions about gendered difference in the ways in which male and female poets perceive, understand and write about the non-human natural world and to further obfuscate the network of mutual creative influences so often ignored by narratives of poetic genius.

While the work of second-wave feminist critics and anthologists was, thus, crucial in putting women writers of the Romantic period on the map of twentieth-century literary criticism in the first place, the underlying conflations of perceived gender and socially constructed gender rules that often fuel their arguments do not only reiterate and, ultimately, often celebrate, the male/female, culture/nature binaries that originally led to the exclusion of women writers based on their perceived 'minor' status and talent; but they also reflect an understanding of "nature" that is fundamentally anthropocentric and binary. Mellor acknowledges this difficulty, arguing that "this binary structure has the initial [...] advantage of allowing us to *see* what has hitherto been hidden, the difference that gender makes in the construction of British Romantic literature". Such a model, however, has its limitations, as it is "theoretically dubious and critically confining. As the obsession of the male Romantic poets with the principle of polarity might indicate [...] a binary model is already deeply implicated in 'masculine' Romanticism."[124] Instead, she argues, "the relationship between 'masculine' and 'feminine' Romanticism is finally not one of structural opposition but rather of intersection along a fluid continuum."[125] Mellor's argument that "[a]ny writer, male or female, could occupy the 'masculine' or the 'feminine' ideological or subject position, even within the same work",[126] suggests that constructions of gender in Romantic writing are more complicated than the binary distinction into male and female writers suggests. Duncan Wu (in his introduction to the hugely influential *Romanticism* anthology) similarly warns against using gender as a distinctive poetic criterion: "Were one to argue that it determines the selection of subject-matter, genre, and poetic form, it would be necessary also to prove that certain subjects, genres, and forms were chosen only, or at least predominantly, by women."[127] While, "[n]o doubt women brought to their art insights

---

124 Mellor, *Romanticism and Gender*, 3.
125 Mellor, 3.
126 Mellor, 3.
127 Wu, *Romanticism*, xxxv.

deriving from gendered experience, [...] those distinctive elements need to be seen in their larger context as vectors in the personality of each writer."[128]

As Mellor argues, questions of whose experiences of "nature" might have been valued are also inextricably entangled with genre as well as gender,[129] as "Romantic women writers challenged the masculine 'feminization of discourse'" in the Romantic novel and "[w]omen poets engaged in ideological contestation, one that affected their choice of both subject and poetic form."[130] Mellor's and Wu's arguments draw attention to the prevalence as well as the oversimplification of binary constructions of gender and genre. In highlighting the strategies of women writers and comparing them to the work of canonized male writers, my aim does not lie in further cementing these problematic binaries but rather in tracing critical narratives of the Romantic poet as solitary genius which tend to obscure diverse strategies and connections and further cement the intrinsic separation of "man" and "nature" that complicates ecological thought in the anthropocene. In my readings of three texts by women writers of the Romantic period in the last part of this chapter I will highlight some of the strategies employed by women writers to (re-)negotiate a Romantic subject position for themselves and to establish their own relationship with an already fraught concept of "nature" in the context of climate crisis.

Women Writing "Nature" and Climate Change in the Early Anthropocene

In the previous parts of this chapter, I have outlined two distinct critical narratives around the work of Romantic women writers and its exclusion from the Romantic canon. While, as I have argued, mainstream Romantic criticism has historically been interested in processes of conceptual unification and canon formation, recent efforts, predominantly by feminist critics, have attempted to undermine critical narratives that excluded women based on their divergence from these unified standards. These two critical approaches are, as I have argued above, distinguished by the ways in which they understand, frame and reiterate the Romantics' specific construction of central concepts like "nature" and the ways in which the poetic subject relates to this entity. Mainstream Romantic criticism, drawing on the canonized Romantics' own conceptualization of poetic genius and a model of "nature" aesthetics inherited from the Burkean sublime, constructed the poet's relationship to "nature" as one of mastery from an external subject position. As I have argued, this has far-reaching consequences based on the pervasiveness of these constructions throughout the anthropocene, as the current popular understanding of ecological concerns

128 Wu, xxxvi.

129 see Mellor, *Romanticism and Gender*, 6.

130 Mellor, 10.

continues to locate "the human" outside of a distinct, externalized understanding of "nature".

I have also outlined the work of predominantly feminist critics of women's writing of the Romantic period, which has attempted to put the work of women writers back on the critical agenda by challenging the dominant critical narrative and by drawing attention to the ways in which women writers' engagements with "nature" complicate the understanding of "man"'s relationship to "nature" put forth by mainstream Romantic criticism. In this last part of the chapter, I will present readings of three different texts by women writers of the Romantic period to outline some of the strategies women writers employ to engage with "nature". I am particularly interested in the ways in which women writers present emergent ideas of climate change and climate catastrophe in the early anthropocene and the narrative and poetic strategies they employ to negotiate a sense of precarity and fundamental change.[131] These strategies include the construction of distinct subject positions, both in poetry and in narrative text, as well as a focus on community and complex perspectivity that I will highlight in an excerpt from Mary Shelley's novel *The Last Man* below.

Communities and connections are central concerns for women writers of the Romantic period, emerging in their writing in multiple, often meta-textual ways that engage with Romantic conceptualizations of the poet as isolated genius. As Hay argues "[t]he myth of the isolated artist has had profound cultural significance over the past two centuries."[132] Romantic poetic manifestos such as Wordsworth's preface to *Lyrical Ballads* (1800) and Percy Shelley's "Defence of Poetry" shaped a historical view of

> the way we think about creativity and genius. Creativity was repositioned as something internal and personal, and poetry – despite its political significance – as the product of an individual's communion with his own mind. The artist became an isolated figure, striving alone to create works of genius.[133]

Instead, Hay's own biographical work on the Byron-Shelley circle in *Young Romantics* (2010) highlights the connections and entanglements of "a group in which friendship was politically and philosophically significant",[134] ultimately shaping and impacting the work of the writers in many ways. To omit the contributions of the women who

131 This project is not so much focused on the ecopoetics of women's writing of the Romantic period, but on the question of the ways in which their writing on climate crisis complicates a critical narrative of Romantic ecopoetics. (see Hume and Osborne, *Ecopoetics: Essays in the Field*.)

132 Hay, *Young Romantics*, xvii.

133 Hay, xviii.

134 Hay, xviii.

formed part of this group based on their gender and to read their work as separate and fundamentally different from that of their male contemporaries, thus, seems counterproductive to understanding both. Moreover, as Hay shows for the Byron-Shelley-circle, mentorship and mutual support had a strong impact on the rise to fame and later canonization of the Romantics guaranteeing in many instances that work was published at all or cementing its literary success.[135]

What also emerges from Hay's account, however, is an additional layer to the work of women writers engaged in these literary circles, as their supportive editing and critical work often obliterated their own efforts, as well as their engagement in the wider context of Romantic poetics. As Hay shows for Elizabeth Kent – entangled with the Byron-Shelley circle as Leigh Hunt's sister-in-law and, at times, collaborator – gender as well as class played a role in how perceptions of "nature" could be framed and written about and the ways in which these questions are further connected to questions of genre. Published in 1823 and now still frequently ascribed to Leigh Hunt as co-author, Kent's *Flora Domestica* mingles instructions "about how to look after pot-plants" with wider information about plants and "quotations from English and Italian poets, from the Greeks and from living writers."[136] As Hay argues, the text "showed that the beauties of nature were available to all" and that "[o]ne did not need to be rich enough to travel to experience the pleasures of nature, since nature could be domesticated in a portable garden."[137] In the climate of literary elitism upheld by the later Wordsworth and his followers (whose arguments about the necessity to restrict access to the English countryside I will discuss in more depth in my chapter on "Steam" below), this was a revolutionary gesture as it directly countermanded the harsh criticisms voiced against the "Cockney School" (including John Keats' *Endymion*) by John Lockhart in *Blackwood's Magazine*. As Hay argues, Lockhart did not only criticize the "moral depravity"[138] of Hunt's circle, explicitly conflating their personal lives and public personas, but he also "stereotyped the Cockneys as suburban gardeners, and attacked them for their limited knowledge of the beauties of nature untamed." As Hay shows, "Bess [Kent] took this stereotype and subverted it, proclaiming a message of democratic luxury and suburban pleasure in the process."[139]

135 see Hay, 58; 99.

136 Kent, *Flora Domestica*.

137 Hay, *Young Romantics*, 285.

138 Hay, 134.

139 Hay, 286.

Elizabeth Kent: Flora Domestica

Encompassing an impressive number of domestic and foreign species, Kent's "Directions for the Treatment of Plants in Pots" describe how to care for an alphabetized and cross-referenced list of plants from "Adonis" and "African Lily Agapanthus" to "Zinnia" and "Zygophy'llum". Moreover, the introduction also proclaims a politics of care that extends beyond the human:

> As I reside in town, and am known among my friends as a lover of the country, it has often happened that one or the other of them would bring me consolation in the shape of a Myrtle, a Geranium, an Hydrangea, or a Rose-tree, &c. Liking plants, and loving my friends, I have earnestly desired to preserve these kind gifts; hut [sic] utterly ignorant of their wants and habits, I have seen my plants die one after the other, rather from attention ill-directed than from the want of it. I have many times seen others in the same situation as myself and found it a common thing, upon the arrival of a new plant, to hear its owner say, "Now, I should like to know how I am to treat this? Should it stay within doors, or without? Should it have much water, or little? Should it stand in the sun, or in the shade?" Even Myrtles and Geraniums, commonly as they are seen in flower-stands, balconies, &c., often meet with an . [sic] untimely death from the ignorance of their nurses. Many a plant have I destroyed, like a fond and mistaken mother, by an inexperienced tenderness; until, in pity to these vegetable nurslings and their nurses, I resolved to obtain and to communicate such information as should be requisite for the rearing and persevering a portable garden in pots. This little volume is the result; the information contained in it has been carefully collected from the best authorities; and henceforward the death of any plant, owing to the carelessness or ignorance of its nurse, shall be brought in, at the best, as plant-slaughter.[140]

I have quoted this opening passage in its entirety, to show how it weaves a network of care that includes beloved "friends" whose gifts of living things are valued both aesthetically and emotionally and "plant-nurslings", whose position is compared to that of children and whose death is likened to "slaughter". There are several strategies at work here that fit stereotypes of women's writing in the Romantic period, from a focus on domesticity, care-work and the foregrounding of relationships, to an emotional connection to beautiful elements of an also feminized "nature". All of these can, however, also be reframed as strategies of engaging with and writing about "nature", beyond the necessity of "mastering" it. In this, they prefigure current feminist discourses around climate mitigation strategies that engage with ideas such as care and co-habitation with non-human animals and other species on this planet.

---

140 Kent, *Flora Domestica*, xiii–xiv.

Moreover, while Kent promises to provide the requisite information to prevent unnecessary plant-deaths, she does not claim a position of authority or mastery over "nature", but instead presents herself as a fellow student of plant life, who has "obtained" and "carefully collected" information "from the best authorities", that she is now able to share with her readers. While this gesture might easily be framed (and gendered) as humility and a failure to claim authority over her own work, it also masks how much work collecting, collating and providing access to this information would likely have required, and downplays the fact that she is presenting a significant amount of botanical information to a readership who would likely not have had access to it from other sources. Potted plants and suburban gardening, were and are, of course, seen as the opposite of wild "nature", or perhaps its already mastered and domesticated side that could, safely, be left to women to care for precisely because it could not be aestheticized beyond the beautiful and could, thus, not offer an engagement with the sublime to the autonomous self. Such an argument, however, disregards the constructedness of both the "wild" and the "domesticated" version of "nature". In highlighting her understanding of caring for "nature" and her own strategies of presenting plant knowledge, Kent's work also functions as an example of the ways in which women writers' engagements with the non-human natural world might have been misread as attempts at celebrating a domestic idyll instead of entries into a broader cultural discussion of writing about "nature" and who could engage in it. As these critical framings are often also tied to discussions of genre, I will further highlight the ways in which ideas of positionality, community and the relationship around "nature" emerge in narrative and poetic examples below.

Mary Shelley: *The Last Man*

My second example comes from Mary Shelley's post-apocalyptic novel *The Last Man* (1826), which I will discuss in more detail in the chapter on "Apocalypse" below. The passage I would like to draw on here is a particularly interesting example of the novel's foregrounding of the relationship between "man" and "nature" as it subverts an understanding of mastery by taking "man" out of the picture. The novel is a pandemic narrative, presenting a long-drawn process of eradication and exile as the survivors attempt to find their space in a progressively empty world. The passage below describes Verney's – the male narrator's – parting look at the deserted English countryside before he attempts to join a group of emigrants who are leaving the country for a warmer climate:

> For the last time we looked on the wide extent of country visible from the terrace, and saw the last rays of the sun tinge the dark masses of wood variegated by autumnal tints; the uncultivated fields and smokeless cottages lay in shadow below; the Thames wound through the wide plain, [...] the cawing of the myriad

> rooks which inhabited the trees of the little park, as in column or thick wedge they speeded to their nests, disturbed the silence of evening. Nature was the same, as when she was the kind mother of the human race; now childless and forlorn, her fertility was a mockery, her loveliness a mask for deformity. Why should the breeze gently stir the tree, man felt not its refreshment? Why did dark night adorn herself with stars – man saw them not? Why are there or [sic] fruits, or flowers, or streams, man is not here to enjoy them?[141]

In this passage from chapter 22 of the novel, Verney's perspective is shored up by the silent presence of his wife, Idris, one of the descendants of the former royal family. While Idris' view is not immediately reported, her presence is included in the passage which describes how "we looked on the wide extent of country" visible from a specific vantage point (the terrace) before it moves on to contemplating the relationship between "nature" and "the human race". This "we" already undermines the dominant Romantic subject position of the autonomous self, as it suggests a shared perspective and view and appears as distinct from "man" evoked as a representative of "the human race" in the latter part of the passage. While the individual, situated view, thus, triggers philosophical thoughts, the narrator does not explicitly include himself or his companion in the generalized category of "man", opening a third, observing position that flanks the gendered binary of "nature […] the kind mother of the human race" and "man". This perspective is frequently reiterated throughout the text as Verney, the eponymous last man, often seems to exist on a different temporal level as an observer describing the novel's main events from a retrospective position.

This temporal distance is also visible in the view of the countryside, which is described as something that used to exist in a specific way, rather than a view in the here and now: "uncultivated fields" and "smokeless cottages" stand as a testament to past efforts at cultivating the landscape in the service of its inhabitants, making "a mockery" out of "nature's fertility", which can no longer be used by mankind. While this seems to be a utilitarian position bent on the exploitation of natural resources, the argument needs to be read within the narrator's glorification of the idyllic, rural past that dominates his narrative of a distant future in the late twenty-first century, unaffected by industrial efforts. Verney's narrative, thus, describes the type of "haunted landscape of the Anthropocene" discussed by Gan, Tsing, Swanson and Bubandt[142] – a now abandoned space bearing the indelible traces of cultivation/exploitation.[143] For Verney the haunting, does not, however, emerge from the traces on the landscape, but from the absence of human life signaled by the landscape, most

141 Shelley, *The Last Man*, 329.

142 Gan et al., "Haunted Landscapes of the Anthropocene," G1.

143 A pure, "unhaunted" landscape is, of course, a construct of Romantic ecological thought in the first place. (see McCarthy, "Beyond Romantic Nature: Ecocriticism's New Shades of Green," 281–82.)

prominently the "smokeless cottages". While the image of smoke is more frequently invoked as a negative, polluting presence in the climate narratives I discuss in this chapter and in the introduction, its absence is presented as a lack of human life in this metonymical construction. Verney's description of the landscape is based on a Romantic understanding of harmonious rural life in an abundant expanse of green and pleasant "nature" that directly and aesthetically benefits the human beings inhabiting it. This understanding of "nature" as a home (associated with motherly nurture), rather than a wilderness to be conquered resonates with the current romanticized understanding of "nature" as predominantly green and preserved for the aesthetic pleasure of humans that I have outlined above. As White argues, this understanding of British landscape as "green" is modeled on a Wordsworthian ideal that presents "nature" as an aestheticized experience of a rural past.[144]

While the novel presents different versions of the non-human natural world, the "nature" evoked in this passage is intrinsically linked with the idea of human cultivation and continuous, unchanging presence, now undermined by the absence of its former inhabitants. In this view, loveliness and fertility go hand in hand, as they are parts of "nature's" aesthetic and nurture of her human children. The passage conflates different forms of nourishment – showing both aesthetic experiences and means of survival in its attempts at describing elements of "nature" in need of an audience. In this curious construction that makes "nature" dependent on human experience and consumption, she remains "the same" but also now appears "childless and forlorn". This view is supported by the descriptions of elements of the landscape associated with ever more dramatic endings – such as the last rays of sunlight, the autumnal leaves and the rooks cawing in the woods – signaling the end of the day, the end of the year and, finally, the end of mankind – ultimately raising the question whether this should not also result in the end of "nature", or at least this aesthetically constructed "nature" which depends on mutual interaction rather than human mastery. While the language of the passage evokes sublime, rather than beautiful or picturesque imagery – for instance in the "dark masses" of wood and the ominous cawing of the rooks – there is no sense of cathartic sublimation for the observing subject and no mastery of a wild "nature" in the aesthetic experience of the autonomous self. Verney is unable to position himself as the autonomous master of this aesthetic experience, precisely because his aesthetic experiences depend on a sense of community with other human beings as well as a benevolent, motherly "nature". The text presents this as a philosophical conundrum as Verney contemplates

144 see White, "Untidying the Landscape: Romantic Poetics, Class and Non-Human Nature," 231. For a critical discussion of Wordsworth's conceptualization of "nature" beyond the romanticization of twentieth-century ecocritical views see Engh, "Wordsworth, Ecocriticism, and Natural Education."

the reasons for "nature"'s existence beyond its utility for human life in a series of rhetorical questions centering on "man"'s experience of nature's abundance.

The novel provides different answers to these questions, which become more and more despondent after the deaths of Verney's companions. The passage in chapter 22 is made ominous by the novel's retrospective narrative perspective, as it is clear from the start that Idris' immediate assurance of love and companionship can only provide temporary relief from these pressing questions: "And since in the wide world though remainest, why should I complain? Thou and nature are still true to me."[145] The closely related construction of "[t]hou and nature" is no coincidence in this passage, as it draws attention to the precarity of the relationship with one which is dependent on the relationship with the other. In *The Last Man*, the aesthetic experience of "nature" is not framed as a solitary one, even though the passage quoted above seems to suggest it by drawing on "man" as the experiencing subject missing from the equation. This "man", however, functions as a collective term used synonymously with "the human race".

While Verney's concept of "nature" is, therefore, based on humanity's relationship to it, his retrospective contemplation of "nature" is not directed at a solitary aesthetic experience but based in a sense of communality and companionship with others, which is essential to his aesthetic interaction with the non-human natural world. This complex aesthetic relationship is only hinted at in the passage above but becomes clearer throughout the course of his narrative until he ultimately turns from "nature" after the deaths of his companions. Remaining as the last man on earth, Verney no longer finds solace in his surroundings, but instead vows to "not live among the wild scenes of nature, the enemy of all that lives."[146] The difference between the kind, nurturing "mother nature" of the passage above and "nature" as "the enemy of all that lives" is not a change in "nature" but a change in Verney's perception of it, as he sees himself as an outcast who is no longer connected to the non-human natural world by means of a common aesthetic experience of "nature" to be shared with his companions. It is in this sense, that "nature's loveliness" becomes "a mask of deformity" after the death of mankind.[147] This construction runs directly counter to the figure of the autonomous self proclaimed by Romantic criticism, whose solitary aesthetic and intellectual engagement with the non-human natural world continue to echo as pervasive cultural narratives of "man"'s mastery over "nature". By reframing both "man" and "nature" the passage and the novel thus also draw attention to Romantic conceptualizations of the non-human natural world as a human aesthetic construction.

---

145 Shelley, *The Last Man*, 329.

146 Shelley, 460.

147 Shelley, 329.

Letitia Elizabeth Landon: "The Factory"

I will return to a more extensive reading of *The Last Man* in the chapter on "Apocalypse". Here, I would like to take my inquiry into the ways in which women's writing in the Romantic period engages with "nature" and in this case also the literary exploration of climate change to a third and final example: Letitia Elizabeth Landon's poem "The Factory" (1835).[148] While I have presented two different engagements with "nature" in the work of Elizabeth Kent and Mary Shelley in the two previous readings, non-fictional and fictional texts are often framed as especially feminine genres by Romantic criticism – a niche for women writers of the Romantic period to explore their on ideas and experiment with constructions of community and positionality. Landon's text takes these ideas and questions to the Romantic poetic tradition, further complicating them by setting them in the context of early anthropocene climate change.

Landon's career sits on the cusp between Romantic and Victorian writing and is often biographically framed as excessive or "Byronic", based, to an extent, on the wild speculations about her personal life and sudden death at the age of 36.[149] Publishing mainly under the pseudonym "L.E.L.", Landon was a critically acclaimed and popular author, whose writing career supported her family after her father's death, but, as Adriana Craciun argues, the critical discussion of her career has so far mainly focused on "the destructive degree to which Landon herself inhabited the persona of heartbroken, beautiful femininity, and the conflicted status of 'poetess'."[150] Craciun's chapter on Landon in her *Fatal Women of Romanticism* (2002) argues that recent criticism frequently demonstrates a lack of awareness of the ways in which Landon's work itself reflects the complex, embodied situation of the female poet as well as Landon's meta-poetic reflection of "the patriarchal limits placed on the woman poet and her impossible choice between love and fame."[151] As Craciun argues, Landon's work (in *The Faery of the Fountains*) meta-poetically explores "a series of painful exiles, the aesthetic dilemma of the Romantic woman poet who is exiled not once like the male poet, but twice."[152] In the example I will discuss below, this is not so much an exile from "nature" itself, but an exile from a specific understanding of "nature" that is available to the male poet in his attempt at fusing the autonomous self with "nature". The poetics of Landon's later work in particular displays a distrust of "all poetic language and transcendence"[153] which Craciun (using Marjorie Levinson's term)

148 Landon, "The Factory."

149 see Craciun, *Fatal Women of Romanticism*.

150 Craciun, 195. See also my discussion of Ferber's chapter on "The Poetess" above.

151 Craciun, 196.

152 Craciun, 210.

153 Craciun, 196.

contrasts with "Wordsworth's 'great escape'."[154] As Craciun argues, the "rhetorical question [in 'The Prophetess'] about the ultimate impotence of the Wordsworthian 'mighty mind' is one of many such subtle critiques of Wordsworth found throughout Landon's work, a critique that places Landon's poetics at the heart of Romantic-period debates over idealism and [the] role of the poet."[155]

In "The Factory" (1835), Landon uses a meta-poetic approach to address the social injustice of child labor, drawing on the metaphorical construction of life as a form of writing. The poem – a ballad of 23 stanzas – is an example of what Craciun aptly titles Landon's "miasmatic discourse", focused on and shaped by questions of life and death raised by the sanitary crisis of the early decades of the nineteenth century.[156] Addressed to "England", which is invoked as "great and free", the text stages a political intervention by presenting the life of factory workers as "[o]ne long and living tomb".[157] While the text draws on a well-established register of social protest imagery that reaches back to Blake's criticism of England's "dark Satanic mills" in *Milton*[158] and his depiction of child innocence and child labor in the *Songs of Innocence* and *Songs of Experience* (1789), it does not present its verdict from the same kind of authoritative positionality, nor does it provide the possibility of a Wordsworthian escape into "nature". As Craciun argues, "[i]n Landon's work, in contrast with Wordsworth's, nature is neither soothing nor even ultimately accessible. [...] While Landon's nature does influence human subjects, it often does so in an infernal miasmatic inversion of Wordsworth's concept of nature's benevolent and purifying influence."[159] There are several examples of this "infernal nature" in "The Factory", but the most striking one is certainly the metaphorical construction of "[w]eeds darkening on the bitter soil / That never knew a flower", which the poem employs to draw attention to the miserable life of the factory workers.[160]

In "The Factory", "nature" also emerges as a positively connoted but temporally distanced entity in the contrast between past and present – an idealized, idyllic childhood and the "living grave" of child labor. The flowing rhythm of the ballad stanza in regular, alternating iambic tetra- and trimeters supports the poem's rousing message that is, as the reader only eventually discovers, addressed at "England". The poem uses somewhat antiquated but less formal "thou/thy/thine" pronouns instead of the more direct "you" throughout the text and the addressee is only named

---

154 Craciun, 197.

155 Craciun, 200.

156 Craciun, 198.

157 Landon, "The Factory," 215.

158 Blake, "Milton," 120.

159 Craciun, *Fatal Women of Romanticism*, 234.

160 Landon, "The Factory," 214.

in the last stanza. The speaker adopts an observing position, predominantly describing even highly emotional reactions as either attributed to others (for instance, "[h]ow precious is the little one, / Before his mother's sight" in stanza 8) or adopting a stance of general outrage (for instance: "Good God! to think upon a child / That has no childish days" in stanza 19).[161] This distanced stance underlines the political argument, which is ultimately phrased as a pleading question in the penultimate stanza:

Is there no pity from above, –
No mercy in those skies;
Hath then the heart of man no love,
To spare such sacrifice?[162]

The use of "man", which might stand for humanity, is ambivalent here – as is the use of "above", which seems to refer both to "those skies" – likely used instead of the more clearly religiously connoted "heavens" to fit into the predominantly masculine rhyme scheme and regular meter – and those politically or socially "above" and therefore in charge of decision-making processes that might end child labor. In this case, it would be the latter group which seems to be implicated by the question "[h]ath then the heart of man no love [...]?" – rather than mankind in general. This additional, political meaning is achieved by the framing of the previous and following stanza which evoke "many an English home" and "England", proclaimed "great and free" by its "tribute waves", before the poem ends on the somewhat ominous line "[t]here is a curse on thee!"[163] While this last line might, of course, be read as a straightforward statement of fact, connected to the poem's subtitle "'Tis an accursed thing!", the exclamation mark and position of this last line lend it an additional performative aspect, suggesting that the speaker might actually be cursing England, its "idol, Gain" (in stanza 7) and the "accursed" factories that demand the "sacrifice" (stanza 22) of children to idolized commercial interests. There is a further, numinous component to the concept of the "idol", which is addressed in stanza 6 and 7, the only instance in which the speaker overtly identifies with a subject position in the poem:

161 Landon, 214–15.

162 Landon, 215.

163 Landon, 215.

We read of Moloch's sacrifice,
We sicken at the name,
And seem to hear the infant cries –
And yet we do the same; –

And worse – t'was but a moment's pain
The heathen altar gave,
But we give years, – our idol, Gain,
Demands a living grave![164]

The juxtaposition of the horror of biblical legend and the current practice, which is described as "worse" in its ongoing cruelty, achieves a shock effect, that is supported by the immediateness and guilt of the "we" from which neither the speaker nor the reader can escape: "We" are all implicated in this construction as "we do the same" as the misled disciples of Moloch, sacrificing children's lives in the service of our "idol, Gain". A use of "you" would have been both more accusatory and easier to refute or protest against than this universalizing "we". The speaker's inclusion in this group, however, also posits an admission to being a beneficiary of this capitalized "Gain" – if not a member of the political entity of "England", which is addressed as "thee" throughout. This complicated subject position allows the speaker to present two versions of mother and child from an external perspective that hinges on an idealized image of "childhood's lovely time" as "innocent and free" (stanza 10 and 11) and the harsh life of the children who "[d]roop, sicken, pine, and die" in the factories.[165] While the poem uses the bond between mother and child to engage the readers sympathy, the speaker is positioned at an observing and commenting distance, drawing attention to how "childhood [should] ever be" (in stanza 12) and asking the reader to "[l]ook on yon child" (stanza 15) as an example.[166]

The poem's imagery relies on binary oppositions (life/death, childhood/manhood), supported by metaphorical constructions that draw on a stereotypical view of beautiful and harsh natural surroundings (light/dark, summer/winter, night/day). The innocence and the memory of childhood are compared to "the freshness of […] spring" (in stanza 12), whereas the child laborer must (unnaturally) rise before daylight and walk to the factory "pale and cold", while his mother's "spare cloak's scanty fold / Scarce screens her from the snowy north".[167] The repetition of sibilants underlines the impression of harsh, hissing winter winds in this line, drawing attention to the poem's careful construction of its imagery. The most interesting passages of the poem in this respect and for my discussion here occur in the first four stanzas,

164 Landon, 213–14.
165 Landon, 214–15.
166 Landon, 214.
167 Landon, 215.

which introduce the factory, before stanza 5 turns to the "low appealing cry" of "[a] thousand children [...] resigned / To sicken and to die!" and, thus, the central message of the text.[168]

The poem opens on the ominous image of "a shade above yon town", creating both a localized threat (yon) and a distanced position for the speaker/observer. The imagery of the following lines oscillates between signs of death – "a dark funereal shroud" and a number of natural phenomena that are ruled out as the cause of the ominous "shade": "'Tis not a tempest hurrying down, / 'Tis not a summer cloud." Instead, stanza 2 announces that "[t]he smoke that rises on the air / Is as a type and sign; / A shadow flung by the despair / Within those streets of thine."[169] This is an interesting construction, as the semicolon at the end of the second line allows the reader to pause and contemplate what this "sign" might mean, before the last two lines of the stanza give a symbolic answer: the smoke is a "shadow" of the "despair" in this place. There is no explanation of the possessive "thine" in this context, as the addressee of the poem is only identified in its last stanza, suggesting that the reader might be addressed as part of a collective "thee" of inhabitants of "yon town" and thus in general and not direct possession of "those streets of thine". The speaker's distance is particularly pronounced here, as "yon" and "those" also signal a spatial and emotional removal from the location of "despair". Stanza 4 repeats this gesture, referring to "thy daily life" and "the moral atmosphere / Around" it, which is "[h]eavy with care, and pale with fear / With future tumult rife."[170] The emotions of "despair", "care" and "fear" are not assigned to the addressee here, but to an as yet unidentified group of people, whose plight – and potential for "future tumult" the speaker is attempting to draw attention to in this curious combined gesture of trying to arouse pity and warning of a threat.[171] The ambiguous concept of a "moral atmosphere" also serves a double purpose, as it connects the "nature" imagery in stanza 3 with the emotional and social implications of the looming shadow of factory smoke. While stanza 4 is, thus, focused on human nature and the effects of the factory upon it, stanza 3 describes the changes wrought on non-human nature by the dark cloud of "smoke":

168 Landon, 213.

169 Landon, 213.

170 Landon, 213.

171 The idea of an uprising of workers – framed as "tumult" here – likely draws on the historical context of the Peterloo massacre of 1819, as well as the many local strikes and uprisings that accompanied the early fight for worker's rights in the nineteenth century. The "moral atmosphere" in the same stanza might, however, also be employed in a reading that imagines uprisings and protests over climate.

That smoke shuts out the cheerful day
The sunset's purple hues,
The moonlight's pure and tranquil ray,
The morning's pearly dews.[172]

There are two conflicting systems of temporality at work in this stanza, as the ever-present smoke cloud overshadows the light of sun and moon from day to evening, to night and morning again, creating a sense of permanence for the monotonous "shade" of the factory smoke that overrules the cyclical day-night rhythm of "nature". The factory does not only disrupt the child laborer's natural sleep pattern, as he must rise before dawn, but it also changes the natural rhythms of the town's surrounding landscape, which is continually covered by "[a] dark funereal shroud". The poem, thus, equates the untimely death of the child laborers with the death of "nature" itself in an attempt at presenting child labor as unnatural – they are "[w]eeds darkening on the bitter soil / That never knew a flower."[173] While this metaphorical construction seems, somewhat harshly, to compare the child laborers to (unwanted) "weeds", the position of this line in stanza 14 highlights its shock value, as it occurs after and in opposition to the description of an idyllic childhood in stanzas 8 to 10. In this context, the non-human natural world functions as both a source of imagery (for the positive depiction of childhood) and an image of "normal" nature which is obliterated by the dark smoke-clouds of the factory. Line 13 begins on a localization: "But here the order is reversed", anchoring the problem of child labor in the present time and current location of the poem, where "infancy, like age, / Knows of existence but its worst."[174] The comparison to "weeds", thus, occurs as part of a description of the working class, forced to suffer under these conditions. This construction is underlined by the meta-textual image of life as writing – "[o]ne dull and darkened page; – / Written with tears, and stamped with toil" – which draws attention to the brevity and monotony of the lives it describes, as well as the crushing sense of institutionalized exploitation expressed by the idea of being "stamped with toil". In this meta-textual gesture, the poem draws attention to the fact that its signaling occurs on many levels, from the smoke cloud as a sign of despair to the short lives of the workers in this "living tomb".[175] These allow the speaker to withdraw behind the signage of imagery and depersonalize the poem's claims. The use of the ballad stanza and the meta-textual image of life as writing – which, in turn, highlights the two versions of childhood the poem presents as exemplary narratives – create a sense of universality based on observation, not sublimation. For Landon, human beings and

172 Landon, "The Factory," 213.
173 Landon, 214.
174 Landon, 214.
175 Landon, 215.

"nature" are connected in a Rousseauean innocent state of childhood, and "man" can draw on this "fairy well; to bring / To life's worn, weary memory / The freshness of its spring."[176] Unless it is interrupted by the unnatural demand to be a "Man from the cradle" and "earn their daily bread", this connection to "nature" is always present, relying on memory, rather than a process of sublimation or mastery.

The poem presents a traditional understanding of gender roles in the context of labor. While in an ideal childhood the child is "innocent and free;" and "[h]is mind essays its early powers / Beside his mother's knee" in a "dear home" that "he will think of" in the "after-years of trouble come", the child laborer is deprived of this early experience as well as any sense of mental development, as "the heat and toil of noon" are "heap[ed] on an infant's head" who, then, knows only one state of existence – "its worst". These two images are connected by the mother's awareness of the injustice from which her "eyes / Turn mournfully away."[177] While labor is seen as masculine – the child laborer is forced to become "[m]an from the cradle" – care is feminized, but remains powerless, as the mother accompanies the child laborer to work in the early morning – presumably to work as well, although the poem does not mention women's involvement in the work force. The system of factory work, and specifically child labor, thus, disrupts traditional gender roles in that the mother is deprived of her caring role and must instead "mournfully" enforce the necessity to get up and work. The disruptive force of this injustice is underlined by the harmonious relationship with "nature" in the ideal childhood, which does no longer exists in the world of child labor. Summer and light – the "precious" child's "hair dancing in the sun" as he is adoringly observed by his mother – is juxtaposed with the "snowy north" wind and winter darkness they must face as she "hurries forth" to the factory before daybreak.[178] While the poem's romanticized depiction of a "natural" childhood is used to underline its argument against child labor, the poem's attention to detail in its descriptions of both ideal and horrific states of "nature" draws attention to Landon's interest in the materiality of the world she describes.

For Landon, there is a moral dimension to the connection between humans and non-human nature, which the poem does not only present with regard to its call to address the injustice of "small children pin[ing] like slaves",[179] but that it also condemns as a *climate* of injustice – the "moral atmosphere" connecting the cloud of factory smoke to the "shadow flung by the despair within those streets".[180] While the

176 Landon, 214.
177 Landon, 214–15.
178 Landon, 215.
179 Landon, 215. A decolonial reading of the last stanza would have to look at the rhyming double of "waves" and "slaves" in the context of "great and free" England, colonialism and the slave trade, but this is beyond the scope of my discussion here.
180 Landon, 213.

poem presents this congruence of moral and meteorological climate as a parallel, rather than a causal connection, the underlying argument connecting human and non-human life and embodiment underpins these parallel narratives of human suffering and climate change. Craciun highlights the "stubborn insistence on the material, corporeal, and historical specificity of literary production" in Landon's work, which she reads as "a significant political aspect of Landon's poetics that has been neglected" by literary criticism so far.[181] While the poem's main focus lies on the "moral" and not the "atmosphere" in this construction, it nevertheless functions as an early example of Romantic literature showing awareness of anthropogenic climate change and its predominantly capitalist cause. Its description of the "shade" thrown by the "smoke that rises on the air" is too specific to dismiss it as merely ornamental or an argumentative sideline supportive of the poem's main cause. Instead, the poem presents itself as an early anthropocene text, that makes an anti-capitalist argument in the service of protecting an interwoven network of human and non-human life against the destructive system of factory labor.

Landon's description of a "shade" connected to the smoke clouds emitted by factories predates Ruskin's observations on the "Storm Clouds of the Nineteenth Century", discussed above. In the lecture Ruskin cites his own diary entry of 1871:

> For the sky is covered with gray cloud; – not rain-cloud, but a dry black veil which no ray of sunshine can pierce; partly diffused in mist, feeble mist, enough to make distant objects unintelligible, yet without any substance, or wreathing, or color of its own. [...] It looks partly as if it were made of poisonous smoke; very possibly it may be: there are at least two hundred furnace chimneys in a square of two miles on every side of me. But mere smoke would not blow to and fro in that wild way. It looks more to me as if it were made of dead men's souls.[182]

Ruskin's descriptions – which I have not cited in full here – are, of course, much more detailed than the few lines dedicated to the shadowy smoke in Landon's poem, and the clouds he describes are always accompanied by stormy winds, which he also describes at length. What interests me here is the parallel Ruskin draws in attempting to determine the cause of the clouds when he first encounters them and describes them in his diary entry, which moves from the image of "a dry black veil" to the idea of "poisonous smoke" from "furnace chimneys" to the imagined substance of "dead men's souls", seemingly building a similar moral argument as Landon's poem. Ruskin was, in all likelihood, not aware of Landon's poem, as he suggests that he could not find any evidence of such clouds in the literature he consulted: "In modern times, Scott, Wordsworth, and Byron are alike unconscious of them; [...]. Taking up

181 Craciun, *Fatal Women of Romanticism*, 198.

182 Ruskin, "The Storm-Cloud of the Nineteenth Century," 3–4.

the traditions of air from the year before Scott's death, I am able, by my own constant and close observation, to certify you that in the forty following years (1831 to 1871 approximately – for the phenomena in question came on gradually) – no such clouds as these are, and are now often for months without intermission, were ever seen in the skies of England, France or Italy."[183] It would be tempting to read Ruskin's move from consulting Scott, Wordsworth and Byron as "modern" writers to "no such clouds [...] were ever seen" as a deliberate omission of the work of women writers, but such an argument would involve further research of Ruskin's reading and observation practices which would move far beyond the purposes of my discussion here. Moreover, Landon's purpose in "The Factory" is a moral argument against industrial processes that involved child labor, not a detailed description of the changes in climate which resulted from these industrial processes.

The ending of Ruskin's lecture is more fatalistic than Landon's anti-capitalist stance on child labor driven by the "idol" of "Gain". Instead of systemic change or a call to collectively oppose the problem, Ruskin suggests that nothing can be done about the weather and, whether it is to be read as a bad omen or not, his fellow citizens should return to their individual "paths of rectitude and piety".[184] What leads me to suggest a comparison between these two texts is, therefore, not the similarity of their arguments, but the parallel structure of their presentation and the imagery they use to illustrate them, as it suggests a connection of climate and ethics – a literal "moral atmosphere" – in early anthropocene literature. Landon's argument moves beyond Ruskin's in its implication of anthropogenic climate change and its association with the exploitation of human life and the overshadowing of "nature" all in the service of material "Gain". The fact that Landon's poem – just as the work of the other women writers I discuss throughout – has not yet seen any critical attention in the context of climate literature underscores the separation of critical approaches to gender and climate in Romantic criticism that I have discussed in this chapter.

Conclusion

In this first, predominantly meta-critical chapter I have attempted to trace the discursive feedback loops between Romantic literature and Romantic criticism which have an impact on current anthropocene discourses, specifically regarding the ways in which the focus on a particular version of Romantic literature highlighted and canonized by Romantic criticism resonates with current discourses around "nature" and the human subject's understanding of and relationship with it. I have described

183 Ruskin, 1.

184 Ruskin, 8.

the processes and strategies used by Romantic criticism to exclude women writers, such as the instrumentalization of the perceived gender of the author and its conflation with thematic and aesthetic strategies. Another, perhaps even more crucial critical argument concerned the identification with self-imposed or externally given labels as a driving factor behind the exclusion of women writers from canonization processes. As I have argued, these processes of inclusion and exclusion cement a specific understanding of the ways in which "man"'s relationship to "nature" emerges in Romantic writing, as well as such related concepts as poetic genius and the autonomous self. In tracing feminist critics' historical attempts at putting the work of women writers on the critical agenda, I have summarized counterstrategies to these processes of exclusion, but also highlighted essentialist arguments emerging from second-wave feminist attempts at conflating gender and embodiment. As I have argued feminist criticism of women's writing of the Romantic period highlights the ways in which these texts challenge and complicate critical narratives of the Romantics' engagements with "nature", as well as some of the counterstrategies that emerge in women's writing on the sublime (such as community or meta-poetic commentary).

In the last part of this chapter I have attempted to trace some of these strategies, most prominently the construction of a subject position in relation to "nature" in three different textual genres to show the ways in which the work of three individual women writers – Elizabeth Kent, Mary Shelley and Letitia Elizabeth Landon – engaged with shifting ideas around "nature", humanity's role in it, community and the possibility of a subject position to engage with "nature" and, in Landon's case, the emerging threat of climate change. While Kent's work might be easily dismissed as amateurish or imitative, the introduction to her *Flora Domestica* presents an ethics of care for human and non-human beings from a non-authoritative position of communal learning. The text itself offers a democratization of specialized botanical knowledge that bridges the perceived gap between country and city, indoor and outdoor "nature". Mary Shelley's *The Last Man*, which I will return to in the chapter on "Apocalypse", poses important questions around the relationship between "man" (or humanity) and a domestic, motherly "nature", drawing attention to the impossibility and futility of a mastery over "nature" in the absence of both "wild nature" and a proclaimed autonomous self. Instead, the novel posits a necessary community of aesthetic experience which the protagonist continues to mourn after the deaths of his companions. While Shelley's novel also deals with climate catastrophe, I have not discussed these passages here, but I will return to this aspect in the "Apocalypse" chapter. Landon's "The Factory" presents the first textual example in this chapter that demonstrates an awareness of anthropogenic climate change. The poem's engagement with "nature" and "climate" is complicated by a subject position shifting between a distanced and an engaged speaker, driving an argument against child labor addressed at a whole country. The detailed discussion of an idyllic childhood and its absence for the children working in factories is embedded in a binary juxtaposition of supportive and detrimental "nature" and an understanding of the connection between the impact on the landscape and the lives of the factory workers in the "moral atmosphere" that overshadows industrialized towns and, by extension, the whole country. Landon's meta-textual work also highlights the connection between literature and activism, drawing further attention to the necessity of including women's writing on climate change in critical discussions of early anthropocene literature, or run the risk of reiterating the feedback loops between canonized Romantic writing and Romantic criticism that I have highlighted in this chapter.

I will continue to read gender as a complicating category in Romantic literature on climate change in the next chapter, which focuses on steam-powered travel in the work of William Wordsworth, Henry David Thoreau and Joanna Baillie. As I will argue, early engagements with steam powered travel engage in complex ways with the challenges to conceptualizations of time and space imposed by the railway and the steam-vessel. I will also address the ways in which these shifting conceptions challenged poetic subject positions and a Romantic understanding of the poet's connection to "nature".

Chapter 2: "Steam"

Introduction: Steam

> The end of the world has already occurred. [...] It was April 1784, when James Watt patented the steam engine, an act that commenced the depositing of carbon in Earth's crust – namely, the inception of humanity as a geophysical force on a planetary scale.[1]

The invention and massive adoption of the steam engine marks one of the potential starting points of the anthropocene as a geological epoch – an argument which can be traced back to Crutzen's original coinage of the term.[2] As John Mee observes, this temporal connection has created a network of critical entanglements between the anthropocene and an "invigorated" Romantic criticism that Mee describes as "Romanthropocentric criticism"[3] – based on Ford's term "Romanthropocene",[4] which I discuss in more detail in the "Anthropocene" chapter above.

The arguments of current anthropocene criticism, discussed in the previous chapter, and the Romantics' engagements with rapid technological change bookend the respective limits of a period of belief in inexhaustible resources, progress narratives and the transformative potential of human inventiveness that they question from two different, but nevertheless interconnected perspectives. Romantic criticism, focused on narratives of progress at the beginning of the anthropocene, frequently frames substantial (industrial) interventions into the landscape such as the building of railway lines as acceleration processes driving early anthropocene

1 Morton, *Hyperobjects*, 7.

2 see Crutzen and Stoermer, "The 'Anthropocene.'" See also the previous chapter on "Anthropocene".

3 Mee, "'All That the Most Romantic Imagination Could Have Previously Conceived': Writing an Industrial Revolution, 1795 to 1835," 229.

4 Ford, "The Romanthropocene," 1.

climate change.[5] For the Romantics, these processes also triggered reflections of the relationship between "humans" and "nature" and the position of humanity in a rapidly changing environment. The adoption or rejection of technological "progress" narratives reframed the ways in which the Romantics saw and aesthetically engaged with this relationship as well as their own positionality.

As a meta-critical term, "Romanthropocene criticism" does not only address these historical connections, but it is also interested in "the transits of meaning[s] which shuttle between the terms it joins together",[6] as well as the underlying implications of both terms – most prominently the difficulty of critically framing the discursive connections between a literary period – with all its critical complexities – and a geological term now generally used to describe a state of emergency that is at the same time so extensive, immediate and entangled with the question of what it means to be "human" in the epoch of anthropogenic climate change that it eludes the grasp of the individual.[7] As I argue throughout, these questions are further complicated by gender (as one, often intersectional, category of marginalization among others), as gendered narratives resist or countermand aesthetic or positional universalizations of the human experience. As I show with regard to Joanna Baillie's poem "Address to a Steam-Vessel" in this chapter, women writers of the Romantic period were, in this context, often engaged in addressing slightly different problems than their male contemporaries, focusing on questions of access and the democratization of aesthetic experience. William Wordsworth's protest sonnet "On the Projected Kendal and Windermere Railway", by comparison, emerges as an attempt at universalizing an individual aesthetic experience by calling on a personified and feminized "nature" to defend herself against "rash assault" in the name of protecting the poet's aesthetic experience of the Lake District from the progressive intrusion of the railway and mass tourism.

As I argue, attempts at critically framing the current moment grapple with the legacy of Romantic positionality, as they draw on Romantic framings of humanity's relationship with "nature". In this context, Timothy Morton's statement (quoted above) that the invention of the steam engine heralded the end of the world may come across as exceedingly narrow in its focus on one specific event whose full effect is only visible in retrospect. Morton draws attention to his own uneasiness with this discursive move – arguing that "[w]e can be uncannily precise about the date on which the world ended. Convenience is not readily associated with historiography, nor indeed with geological time. But in this case, it is uncannily clear."[8] The repeti-

5 see Briggle, *Thinking Through Climate Change*, 12.

6 Ford, "The Romanthropocene," 1.

7 See also my discussion of the term "Anthropocene" and its implications in the previous chapter.

8 Morton, *Hyperobjects*, 7.

tion of the adverb "uncannily" does not only suggest an awareness of the improbability of coincidence, but it also frames a singular past event in terms of its capacity to haunt the present on a scale that moves far beyond its original proportions.

In *Fossil Capital* (2016) Andreas Malm describes Morton's argument as "hyperbole",[9] arguing that the moment of patenting the steam engine did not herald a steep rise of CO^2 in the atmosphere, but rather a gradual one, detectable over the following decades. Malm's argument does not question the impact of fossil fuels on climate change but draws attention to the problems that arise when highlighting singular events as far-reaching and inevitable in narratives about the industrial revolution. He retraces historical accounts postulating a necessity behind the adoption of steam engines, based, for instance on the scarcity of space, energy resources, or the limitations of other forms of energy production, such as the water wheel, to highlight the ways in which a lack of complexity in these accounts drives narratives of progress towards a sense of inevitability.[10] These are often also based on an understanding of "human nature" (humanity's "pyromaniac inclination" or inventiveness) that drive a universalized understanding of the anthropocene,[11] and suggest that technological innovation might also create a way out of the anthropocene predicament.[12] As Malm argues, the rise of steam had more to do with a capitalist interest in the economic growth accelerated by steam power, than with a "universal trait of the [human] species" to burn things[13] or human inventiveness in the service of mankind. For Malm, "steam heralded a new union between growth and fossil fuels",[14] as part of what he describes as a "fossil economy", that is "an economy of self-sustaining growth predicated on the growing consumption of fossil fuels, and therefore generating a sustained growth in emissions of carbon dioxide."[15] Malm's central argument about fossil fuel economies posits that "steam arose as a form of power exercised by some people against others."[16] This challenges historical narratives postulating a necessity or even inevitability inherent in fossil fuel economies, because "if some humans introduced steam power *against the explicit resistance of other humans*, then it would be hard to maintain a notion of it as the expression of a species-wide project."[17]

9 Malm, *Fossil Capital*, 29.
10 see Malm, 26.
11 see Malm, 29–31.
12 Malm, 32.
13 Malm, 31.
14 Malm, 29.
15 Malm, 11.
16 Malm, 36.
17 Malm, 36.

While Malm's arguments question the anthropocene aesthetic[18] inherent in historical narratives about the inevitable or even heroic rise of steam power, this also has important ramifications for the ways in which it might be possible to challenge the hopeful inevitability of futuristic narratives based on technosolutionism and human inventiveness, which I will discuss further in my Conclusion. As I show in this chapter, these are not only questions that arise in Romantic literature around steam-powered travel, but ones that also emerge in and through Romanthropocene connections. As Davies argues,

> the present global environmental crisis is an invitation to understand the Industrial Revolution less as an evolutionary breakthrough to indefinitely self-sustaining technological progress, and more as a fork on to a path of carbon- and resource-intensive production that entails still unfinished reorganizations of the living world and climate system.[19]

Romanthropocene criticism draws attention to these connections in both directions, inviting meta-critical reflections of critical framings alongside specifically situated textual readings. It also creates a critical space to reflect the ways in which the Romantics' reactions to these processes in the early anthropocene might create a better understanding of the current moment. In this context, I present an argument based on the need for complexity with regard to different Romantic subject positions, to draw attention to Romantic legacies in the current conceptualization of the human subject's position in the context of climate crisis. What emerges on a meta-critical level from the discrepancies between Morton's and Malm's arguments is not simply a question of historical framing, but ultimately also the problem of how these framings are reappropriated and instrumentalized in different political and critical contexts and who is excluded from these discussions. While both Morton's and Malm's accounts of climate change are ultimately targeted at describing its complexities and exigencies, I see two distinct argumentative movements at work here – one based on the shock effects inherent in universalization and one interested in situating knowledge in historical specificity to challenge universalized historical narratives.

While I will return to these recent anthropocene discourses in my Conclusion, my focus here is directed towards the ways in which these two argumentative movements – universalization and situated knowledge – emerge in early anthropocene literature and their reflection of different ways of engaging with rapid and substantial technological and cultural changes in texts that revolve around steam-powered travel as a relatively recent technological invention and economic "opportunity". As

18 see Hird, "Proliferation, Extinction and an Anthropocene Aesthetic."

19 Davies, "Introduction: Romantic Studies and the 'Shorter Industrial Revolution,'" 195–96.

I argue, these confrontations triggered discursive shifts in the relationship between the experiencing subject and the non-human natural world that had far-reaching consequences. In this context, I contrast Wordsworth's stance against the railway as a universalizing position with Henry David Thoreau's engagement with the ways in which the railway reshaped notions of time as an example of situated knowledge. Joanna Baillie's poem, as I will argue, presents a more complex, ambiguous stance, juxtaposing positive and negative aspects of steam-powered travel from a perspective that is both balanced and distanced.

Romantic arguments against steam power – Wordsworth's and Thoreau's position for instance – often circulate around economic rather than ecological questions, but this shift in focus also has important socio-political ramifications based on the era's pronounced economic imbalances on the basis of class, gender and racialization. As I will argue below, these are often negotiated via questions of access and accessibility, as steam powered mass transportation pitted individual interests against the possibility of democratizing travel and, by extension, the aesthetic experiences available through it. These questions emerge most prominently in Baillie's poem, but I will also trace questions of access around women's engagements with travel in the concluding part of this chapter, addressing positionalities and situated knowledge in women's travel writing, as well as in the accessibility of different versions of "nature".

In the textual examples I discuss below, these questions play out on a localized and historically specific level, but they are also connected to broader questions inherent in the term "anthropocene" – most prominently the difficulty of conceptualizing humanity's relationship with the non-human natural world in a framework of time and space that operates on a planetary scale and across millennia. A sense of acceleration or shifts in the understanding of temporality, thus, emerges as another central dimension of the ways in which early anthropocene literature engages with far-reaching technological change. Thoreau's *Walden*, most prominently, highlights different temporalities that oscillate between attempts at universalization and standardization in his engagement with "railway time".

By tracing individual positions on steam power and its effects, I attempt to situate these texts both historically and contextually within Romantic discussions of what we now perceive as emergent anthropogenic climate change in the context of steam powered industrialization and commercialized transport. As I will argue, the possibility of steam powered travel heralded profound changes in the Romantics' understanding of such fundamental categories as time and space and, by extension, humanity and "nature", often cementing an understanding of non-human nature or Romantic "nature" as a force outside and beyond human beings.

In the first part of this chapter, I will present William Wordsworth's sonnet and letters "On the Projected Kendal and Windermere Railway" as an anti-railway stance which draws on a universalizing aesthetic to present a position of protest against the intrusion of the railway and tourism into a specific landscape. I highlight the ways in which Wordsworth presents an elitist position by arguing against touristic intrusions into the Lake District and by calling on a personified "nature" to defend herself against these intrusions.

Wordsworth's stance will function as a basis for a comparison with Joanna Baillie's "Address to a Steam-Vessel" in the second part of this chapter. I read Baillie's poem as an engagement with steam-powered travel that uses a distanced perspective typical of women's writing in the Romantic period to present an argument for the accessibility of aesthetic experiences facilitated by steam-powered travel, but which also takes an ambiguous stance towards this new technology based on the aesthetic changes it entails.

In the third part of this chapter, I turn to Henry David Thoreau's *Walden* and its engagement with the railway as an example of highly influential early environmentalist writing rooted in an understanding of situated knowledge.[20] I use Thoreau's detailed account of his understanding of the railway as a commercial project to discuss underlying questions of temporality and socio-economic change.

The concluding part of this chapter addresses questions of accessibility and positionality specific to women's writing around experiences of travel and the connected question of access to the non-human natural world.

William Wordsworth "On the Projected Kendal and Windermere Railway"

Baffle the threat, bright Scene, from Orresthead
Given to the pausing traveller's rapturous glance:
Plead for thy peace, thou beautiful romance
Of nature;[21]

In 1844, William Wordsworth expressed his feelings in opposition to the proposed extension of the railway between Kendal and Windermere in an extensive and sharply worded letter to the *Morning Post*,[22] arguing in particular:

20 While Thoreau is more commonly grouped among the American Transcendentalists than seen as a Romantic, there are parallels and connections between the two movements, especially in the relationship to "nature" as a source of inspiration for poetic or personal growth. As Ruland and Bradbury argue, Transcendentalist "thinking was fed by the spirit of Romanticism" (Ruland and Bradbury, *From Puritanism to Postmodernism – A History of American Literature*, 118.).

21 Wordsworth, *Kendal and Windermere Railway – Two Letters Reprinted from the Morning Post*, 4.

22 For a detailed overview of Wordsworth's, sonnet and two letters sent to the *Morning Post*, as well as contemporary responses, see Wells, "Wordsworth and Railways in 1844–1845."

> What can, in truth, be more absurd, than that either rich or poor should be spared the trouble of travelling by the high roads over so short a space, according to their respective means, if the unavoidable consequence must be a great disturbance of the retirement, and in many places a destruction of the beauty of the country, which the parties are come in search of? Would not this be pretty much like the child's cutting up his drum to learn where the sound came from?[23]

Wordsworth's argument pits an interest in the "beauty of the country", which is also associated with the concept of "retirement", against the interest of travelers of different backgrounds to reach said country, suggesting that to facilitate an easier route *into* the country must result in a "disturbance" of its retirement and a "destruction of [its] beauty".[24] The juxtaposition the passage evokes is not the one between country and city, or even between "humans" and "nature", but rather one of scale – evoking more recent discussions of mass tourism and its impact on the environment.[25] The whole passage deals in absolutes (either there is a railway, or there is a beautiful countryside) and the use of the simile in the last sentence only makes sense in this context, as the railway is positioned as the instrument of destruction of an aesthetic experience dependent on its absence.

Wordsworth resistance against the railway draws on an experience of the landscape that is entirely individual, relying on the absence of other travelers, but presented as universal through the use of both aesthetic arguments (in the letter) and Romantic aesthetics (in the sonnet accompanying it). His main argument hinges on the assumption that the countryside would be destroyed for purely touristic purposes; as he argues, "[i]n this district the manufactures are trifling; mines it has none, and its quarries are either wrought out or superseded; the soil is light, and the cultivateable parts of the country are very limited; so that it has little to send out, and little has it also to receive."[26] This enumeration is necessary, because Wordsworth's argument that better access to the aesthetic experience of the countryside would entirely destroy it can only be made if the railway is otherwise useless. His understanding of tourism, by consequence, "precludes the notion of a railway"[27], not only because someone explicitly travelling to the Lake District to tour it, "would not grudge a

23 Wordsworth, *Kendal and Windermere Railway – Two Letters Reprinted from the Morning Post*, 12.

24 A letter to Henry Crabbe Robinson dated 8 December 1844 indicates that Wordsworth was aware of the opposition against his arguments, as well as their futility: "I am not so simple as to think that my paper if published would stop the nuisance, but collaterally it might set opinion right in some quarters." (Wordsworth and Wordsworth, *The Letters of William and Dorothy Wordsworth – The Later Years, Part II*, 634.).

25 see Yoshikawa, *William Wordsworth and Modern Travel – Railways, Motorcars and the Lake District, 1830–1940*.

26 Wordsworth, *Kendal and Windermere Railway – Two Letters Reprinted from the Morning Post*, 6.

27 Wordsworth, 6.

two hours' walk across the skirts of the beautiful country that he was desirous of visiting",[28] but also because tourism itself, a relatively new phenomenon in the 1840s, was not to be imagined as a mass activity demanding railway access to the tourist site.

Time, or as Wordsworth argues, leisure is an intrinsic element of this envisioned experience of the landscape – an experience that is simply not available to those arriving in it by any means of transportation faster than on foot. For Wordsworth, the aesthetic experience of the landscape is, thus, tied to a temporal framework that is incompatible with speedy travel as it relies on slowness as well as a sense of temporal and spatial immersion. Moreover, as John Mulvihill points out, for Wordsworth, "difficulty" is a necessary component of the aesthetic experience: "[a]lthough difficulty is not necessarily a good in itself, a willingness to negotiate necessary difficulty points to an active and critical taste."[29] Mulvihill discusses the question of access emerging from this argument in terms of class and the democratization of access, concluding that Wordsworth foresaw the more contemporary realization that "mass participation can change nature."[30] The insistence on limiting access to the landscape to those who are able to walk considerable distances or traverse different terrains, however, also raises questions of physical accessibility that I will discuss further with regard to Dorothy Wordsworth's disability in the concluding part of this chapter.

Environmental Elitism

William Wordsworth's opinion in the letter to the *Morning Post* and the sonnet accompanying it emerges as fundamentally elitist, positioning a specific experiencing subject as more in tune with the beauty of the landscape and focusing on a lone, rather than a common experience of "nature" as the basis of an aesthetic engagement that would otherwise be impossible. The arguments in the letter attempt to universalize this experience by postulating aesthetic theories as common sense. The letter expresses the view that humans do not so much possess an innate but rather an acquired, educated taste for picturesque landscapes and the lower classes would, therefore, not benefit from easier access to Lake Windermere because of their inability to grasp its beauty. From Bate's early ecocritical perspective, Wordsworth's arguments can be read in the context of a project of conservationism that draws not only on his connection to and poetic understanding of a specific landscape, but also on a sense of value of the non-human natural world: "The Romantic sublime celebrates the power of the human imagination, but it also powerfully conveys a sense

28 Wordsworth, 11.

29 Mulvihill, "Consuming Nature," 321.

30 Mulvihill, 326.

of the insignificance, the smallness, of man." Bate reads this as a "necessary humbling, a first step towards the knowledge that humankind is not self-sufficient. Like all species, we need our ecosystem; but unlike other species we also have the capacity to destroy it."[31]

To read Wordsworth in this ecocritical context emerges as a personal project in Bate's text – a project which he connects back to his own positionality as a reader, critic and hiker encountering specific ecocritical texts and landscapes, but – as Wordsworth's letter – Bate's reading also attempts to generalize a personal experience into an ecocritical argument that postulates a somewhat universalized "we". Moreover, Bate suggests that Wordsworth's poetry might be a useful entry point into "new ways of conceiving of the planet",[32] but he does not acknowledge the inherent anthropocentric universalization in such a move, which would be similar to Wordsworth's attempt at universalizing his own experience of attempting to preserve his view of Windermere.[33]

Wordsworth's letter shifts from a personal view of the landscape as a place of "retirement" and aesthetic inspiration for the poet to a wider framing of the changes to "the country" and "nature" that railway access for the masses might entail. This argumentative move, which is present in the letter, but emerges even more clearly in Wordsworth's sonnet against the railway, hinges on the word "retirement" (signaling both spatial seclusion and a slowed down experience of temporality), which the sonnet uncouples from an experiencing subject. This tension between a personal and a universal argument in Wordsworth's approach will be the focus of my discussion in the following, placing it in the broader context of temporality and positionality around experiences of steam-powered travel that is the main focus of this chapter.

Framed and introduced by the letter to the morning post, Wordsworth's anti-railway sonnet, begins with a rhetorical question that drops the reader into the middle of what seems to be an ongoing debate:

Is then no nook of English ground secure
From rash assault? Schemes of retirement sown
In youth, and 'mid the busy world kept pure
As when their earliest flowers of hope were blown,
Must perish; – how can they this blight endure?
And must he too the ruthless change bemoan
Who scorns a false utilitarian lure

31 Bate, "Romantic Ecology Revisited," 161.

32 Bate, 161.

33 I have already discussed Davies' assessment of Bate's position in the "Anthropocene" chapter above. Here, I would like to draw attention to the general problems that such a universalization of an individual approach runs the risk of assuming an anthropocentric position and romanticizing a specific experience to the detriment of others.

'Mid his paternal fields at random thrown?
Baffle the threat, bright Scene, from Orresthead
Given to the pausing traveller's rapturous glance:
Plead for thy peace, thou beautiful romance
Of nature; and, if human hearts be dead,
Speak, passing winds; ye torrents, with your strong
And constant voice, protest against the wrong.[34]

Italian in form and written in lines of fairly regular iambic pentameter with a few rhythmic breaks for emphasis, the sonnet aligns itself with other popular Wordsworthian sonnets on humanity's relationship with "nature" such as "The world is too much with us". Its formal elements, however, also support its function as a protest text – for instance the dramatic pauses, the rhyming pair of "Orresthead" and "dead" in the sestet or the opening question. As Mulvihill argues, the first lines "evok[e] the parochial alarmism of the Napoleonic War years",[35] and there was, consequently, substantial criticism of Wordsworth's rhetorical conflation of "urban working-class tourists" and invading Napoleonic armies.[36] While Wordsworth's two letters to the morning post on the subject are carefully phrased to dodge a direct involvement in this controversy,[37] the poem presents an emotional stance that claims authority on the subject and attempts to universalize the experience of intrusion via a direct invocation of "nature".

There is no mention of the railway throughout the sonnet, which leaves it to the reader to construct from the context what exactly constitutes the "rash assault" in line 2. The "ruthless change" and "false utilitarian lure" in line 6 and 7 hint at both innovation and a sense of economic progress. While the octave is focused on personal experiences, hinging on the opposite pair of youth and retirement as well as the use of "he too" that signals an outsider to the debate, the sestet widens the scope and addresses "thou beautiful romance / Of nature" directly, calling on the "passing winds" and, indeed, a personified and romanticized "nature" itself to protest the changes.[38]

Here, the perspective shifts, presenting the "pausing traveller" in line 10 from the outside, as "his" (presumably a male traveler is implied) "rapturous glance" is "given" the "bright Scene" of Lake Windermere and its surroundings that can be overlooked from Orrest Head.[39] Paul Smethurst draws attention to the inherent irony of de-

34 Wordsworth, *Kendal and Windermere Railway – Two Letters Reprinted from the Morning Post*, 4.
35 Mulvihill, "Consuming Nature," 306.
36 Mulvihill, 306.
37 Mulvihill, 311–12.
38 Wordsworth, *Kendal and Windermere Railway – Two Letters Reprinted from the Morning Post*, 4.
39 Wordsworth, 4. James Baker Pyne's painting "Lake Windermere from Orrest Head" (see Pyne, "Lake Windermere from Orrest Head.") captures this view in 1849 – that is two years after the train terminus at Windermere opened (in spite of Wordsworth's protests) and travelers from

scribing "nature" in these elevating tones, as Wordsworth's writing (specifically the *Guide to the Lakes*, 1835), "by idealizing and localising definitive examples of natural beauty, actively produced the Lake District as a tourist site. He created a 'taste' for the region in the vulgar sense of sights/sites to be consumed."[40]

The reference to Orrest Head localizes the poem, but the letters and the sonnet attempt to universalize the question of preservation in an attempt to protect this specific landscape. As Mulvihill argues,

> The particular landscape on whose behalf Wordsworth speaks here is not the only beautiful landscape in England, but it is unique in itself and in every one of its constituent features. As such, the issue in question "is not only different from all others, but, in truth, peculiar." It is a matter that, "though seemingly local, is really one in which all persons of taste must be interested."[41]

By turning the argument for the preservation of a particular scenery into a question of universal taste, Wordsworth attempts to place his protest against the intrusion of the railway into his corner of the Lake District into a wider aesthetic discourse. This argumentative move is supported by the poem's construction of a communicative situation that allows a covert speaker to hide behind an appeal to an overt addressee. While the traveler is introduced as a lone figure who has the time to pause and look at the scenery, he is not the speaker or even the central concern of these lines, which are addressed directly at the personified and capitalized "bright Scene" itself. The poem's covert speaker calls on the scenery to "[b]affle the threat", echoing a sentiment about the importance of the aesthetic experience also expressed in the letter.[42]

Universalizing the Experience of "Nature"

While the sonnet calls on "nature" to defend herself, it is interesting that "nature" is not spelled with a capital "N" (as it is, for instance, in the better known "The world is too much with us"), and is addressed indirectly, in the context of "romance", which evokes an aesthetic/literary construction. It is not a universal Romantic "nature" that is being addressed here, but rather a specific, even localized and, at least to an extent, personalized aesthetic experience. Moreover, the phrasing suggests that "nature" exists not in and for itself, but in the shape of a romanticized experience of it, that relies

---

further afield were able to enjoy the walks and views around Windermere. It made Windermere a popular destination for tourists from the industrial towns in the North of England and brought more residents into the area, but, mostly depending on leisure travel, the train company was not a commercial success and later sold to a larger company.

40 Smethurst, "Prescribing Nature: William Wordsworth's Guide Through the Lakes," 173.

41 Mulvihill, "Consuming Nature," 319.

42 Wordsworth, *Kendal and Windermere Railway – Two Letters Reprinted from the Morning Post*, 4.

on the presence of an experiencing subject. As Evan Gottlieb argues, for Wordsworth "the world is knowable, not to mention meaningful, only in so far as it correlates to Wordsworth's mental experience of it."[43] This "correlationism" is coupled, as Gottlieb argues, with a fundamental understanding of "nature" as still rooted in Christianity in spite of the recent discovery of geological strata in Wordsworth's time that pointed to an understanding of earth's temporality that would later be termed "deep time", which had begun to shake a biblical understanding of temporality:

> whatever geology might be discovering about the earth's unpredictable productivity, its ultimate stability could still be counted on as the basis of human life and, by extension, morality. Just as '[t]he Child is Father of the Man', Wordsworth reassures us that, come what may, our environment will sustain us, spiritually as well as physically, just as the speaker's days, ideally, will be 'Bound each to each by natural piety'.[44]

This spiritual anchoring in a greater power, as well as "Wordsworth's conviction that the natural world fundamentally exists in harmony with the human one, if only we can learn to see it rightly"[45], functions as the basic premise of the speaker's appeal to "nature" to preserve itself. Gottlieb describes temporality in Wordsworth's work as "correlationist time" a time frame which "connects natural history to human history in a reassuringly correlationist manner, such that the expectation of human continuity is underwritten by an ultimately anthropocentric faith in humanity's connection with a natural world perceived by us as metastable and enduring."[46] Correlationist temporality runs through and shapes the aesthetic experience of "nature" in Wordsworth's poetry, making it possible for the speaker to appeal to "nature" to protect itself against "rash assault" in the name of a preservable, enduring aesthetic experience of beauty.

The use of the term "romance" evokes a specific poetic form (the "romance", which in English literature usually presents itself in the form of a ballad stanza) that the poem itself (an Italian sonnet) does not fulfill, but the evocation of poetic form nevertheless draws attention to the meta-poetic function of this construction, which bridges the gap between the poetic form of the text and the ecocritical message it transports. The author of the poem, who clearly positions himself and his opinion in the letter, disappears behind this construction – instead urging for an aestheticized "nature" to defend itself and preserve itself for the personalized aesthetic pleasure of the speaker. As Bryan L. Moore argues, "[t]he melding of external nature and inner consciousness culminated in Wordsworth, for whom nature

43 Gottlieb, "Anthropocene Temporalities and British Romantic Poetry," 30.

44 Gottlieb, 29.

45 Gottlieb, 29.

46 Gottlieb, 32.

involves community both in itself and in its relationship to humans."[47] The aesthetic experience of the "beautiful romance of nature" inherently transports this idea of a melding of human consciousness and "nature", making the explicit appearance of the experiencing subject position redundant.

As the letter, which references "a great disturbance of the retirement",[48] lines 2 to 3 of the sonnet also introduce the idea of "schemes of retirement sown in youth", but a slight difference in meaning emerges between the more general sense of retirement inherent in the countryside and the "plans of retirement" that presuppose a subject hatching such plans in its youth.[49] While according to the *OED*, the sense of the word signifying retreat from public office had already been well established by the 1840s,[50] it is unlikely that this encompasses Wordsworth's full meaning here.[51] Another meaning of the term retirement proposed by the *OED* carries more likely connotations: "a secluded or private place; a retreat" for which the dictionary provides several contemporary examples referencing country retreats and gardens.[52] While the use of the word in the sonnet resonates with the speaker's personal response to this "blight" and "rash assault", the use in the letter suggests a more general position not "*my*" but "*the*" retirement and the "destruction of the beauty of the country" that supports this meaning of the country as a "retreat", signaling, most prominently, the absence of human society.[53] As Lucy Newlyn argues, Wordsworth engagement in the environmental cause of keeping the retirement of the country intact was based on personal feelings, as "[n]ot only would the peaceful surroundings at Rydal Mount be spoiled, but on the other side of the Lake District the planned railway would pass close by John Wordsworth's vicarage in Brigham, cutting through his garden."[54] While, according to Wordsworth's letters this seems to have concerned a different branch of the railway line (not from Kendal to Windermere but from Cockermouth to Workington),[55] both were planned at the same time and one can certainly read the line about "paternal fields" as an expression of this personal interest.

47 Moore, *Ecological Literature and the Critique of Anthropocentrism*, 85.

48 Wordsworth, *Kendal and Windermere Railway – Two Letters Reprinted from the Morning Post*, 12.

49 Wordsworth, 4.

50 "Retirement, n."

51 Mulvihill, "Consuming Nature," 305. Wordsworth had been made poet laureate in 1843 and had, thus, definitely not retired from his public position. Moreover, he was actively drawing on the standing of this position in addressing first the editor of the *Morning Post* and later the prime minister, William Gladstone, in letters concerning the matter of the railway.

52 "Retirement, n."

53 Wordsworth, *Kendal and Windermere Railway – Two Letters Reprinted from the Morning Post*, 12.

54 Newlyn, *William and Dorothy Wordsworth*, 290.

55 see Wordsworth and Wordsworth, *The Letters of William and Dorothy Wordsworth – The Later Years, Part II*, 639. While Wordsworth was unsuccessful in stopping the construction of the railway line, he effected the inclusion of a clause in the contract that led to the building of a new vicarage away from the line.

It is the shift from "my" to "the" between the letter and the sonnet which suggests a universalization of the experience of "retirement" that is crucial to Wordsworth's aestheticized protest against the railway. The attempt at universalizing his argument is built on shaky ground, because the experience the sonnet revolves around is a fundamentally personal one and dependent, as he himself argues, on the solitary presence of the experiencing subject.

In an attempt at instrumentalizing "nature" itself for its cause, the text moves away from "human hearts" which might already be "dead" and calls on personified, non-human elements to oppose the plan: "Speak, passing winds; ye torrents, with your strong / And constant voice, protest against the wrong."[56] Beyond the unified "romance of nature" that includes an experiencing subject, there is a sense of non-human agency in these lines that positions the speaker as external to, but complicit with a personified "nature" or scenery, more than with human society.[57] While this shift between the octave and the sestet resonates with a Romantic understanding of "nature" as external to the autonomous self experiencing it, the sonnet's demanding speaker and foregrounding of an aesthetic experience that necessitates an experiencing subject ties the sonnet to an anthropocentric perspective.

The same holds for the sonnet's presentation of temporality. The juxtaposition of "youth" as a time to make plans for retirement and said retirement in later life suggests an awareness of shifts in temporality brought about by the combined threats of the railway and the ensuing or at least assumed intrusions of industries, trades and travelers into the secluded countryside. This understanding of time is, however, fundamentally human and the landscape, although personified, does not seem to possess its own temporality.

In the concluding paragraphs of the letter Wordsworth draws attention to the materialist interests behind expanding the railway network which he explicitly opposes, all while being aware that the "railway power, we know well, will not admit of being materially counteracted by sentiment" and that all he can propose against these arguments is the beauty of the country and its "character of retirement", which should remain "undisfigured" and "unviolated" by those "who are dazzled by the application of physical science to the useful arts, and indiscriminately applaud what they call the spirit of the age as manifested in this way."[58] Wordsworth's project is clearly conservative – in the sense that it attempts to preserve the current state of the landscape in the face of what was seen as technological and industrial progress

56 Wordsworth, *Kendal and Windermere Railway – Two Letters Reprinted from the Morning Post*, 4.

57 Wordsworth's sonnet "The world is too much with us" follows a similar pattern in introducing personified elements of nature, for instance "the sea that bares her bosom to the moon", also creating a link between the speaker and an at least partly feminized "Nature" to stand against a commercially oriented humanity (Wordsworth, "The World Is Too Much with Us," 372.).

58 Wordsworth, *Kendal and Windermere Railway – Two Letters Reprinted from the Morning Post*, 13.

by his contemporaries (in addition to also being politically conservative in its resistance against a prominent progress narrative of the day) but also in its reliance on the understanding of a stable and universalizable relationship between the experiencing subject and the experienced object.

Wordsworth's counterarguments suggest a deeply felt involvement – almost a sense of kinship – with the landscape that extends to its other human inhabitants (the lone travelers arriving on foot and the farmers tilling "paternal fields" while scorning the "false utilitarian lure") but clearly not to large groups of travelers, whose arrival would not only disturb the retirement but also the aesthetic experience of the landscape.[59] The main divisive factor in this argument seems to be social class – even though, at least on the surface, those already inhabiting the landscape are favored over those who have no connection to it – but the underlying argument against the railway hinges on the *speed* at which the travelers reach the Lake District and its landmarks, precluding railway travel on the basis of not just the destruction of the landscape the tracks would cut through, but also based on the idea that only travelers who take their time to traverse the landscape have the aesthetic (and potentially also ethical) capacity to engage with it. Wordsworth's argument, thus, emerges as a retrospective one, privileging a pre-industrial but still anthropocentric experience of the landscape over the "utilitarian lure" of the railway's progress narrative.

Progress and Ruin

To reject Wordsworth's stance as anti-industrialist would, however, be too narrow an understanding. As Davies argues, Wordsworth's "treatment of 'manufacturing industry,' while certainly conservative in orientation, was by no means simply rejectionist."[60] Instead, his criticism of the railway focuses on its intrusion into a specific, localized space at a specific speed detrimental to aesthetic experience, thus reflecting a somewhat conflicted stance towards railways. As Li-Hsin Hsu argues, "Wordsworth's progressive intent seems irreconcilable with his protectionist instinct. Indeed, one senses subtle hesitation in the 1833 sonnet ['Steamboats, Viaducts, and Railways'] of Wordsworth, in which the tension between the 'harsh features' of the railway and 'old poetic feeling' remains not fully resolved."[61] More-

59 One of the engineers contracted to build the railway wrote a long poem in response to and picking up some of the lines from Wordsworth's sonnet arguing that the railway gave travelers access to the beauty of the landscape and "to town-cramp'd souls the power to soar" (Heald, "Reply to Wordsworth."). Interestingly, Heald's poem echoes many of the ideas of Joanna Baillie's "Address to a Steam-vessel" which I discuss below – including a reference to steamers on Lake Windermere, which might offend "the poet" just as much as the railway.

60 Davies, "Introduction: Romantic Studies and the 'Shorter Industrial Revolution,'" 189.

61 Hsu, "The Romance of Transportation in Wordsworth, Emerson, De Quincey, and Dickinson," 50.

over, a letter to the banker Charles Lloyd written in 1825 and asking Lloyd for advice on which of the railway companies operating out of Birmingham to invest 500 pounds in[62] and an 1840 letter to Cordelia Marshall lauding a train journey Dora Wordsworth took from Preston[63] to London as "most agreeable and unfatiguing"[64] show that Wordsworth was, at least at that moment, not opposed to the idea of railway travel in general.

As Hsu argues, Wordsworth's 1833 sonnet "Steamboats, Viaducts, and Railways", while proclaiming a war between "nature" and technology in its octave, ends on "a poetic (as well as philosophical) attempt to resolve the opposition between the mechanical and the natural" ending on "a cheerful celebration of speed, motion and progress."[65] By 1844 Wordsworth's tone had changed to also include a hint of anti-capitalism alongside his conservationist stance: in a letter to General Charles William Pasley, Wordsworth railed against "the excesses to which the Railway Mania drives people on the present superabundances of Capital", arguing that "the staple of this country is its beauty and that will be destroyed by such a nuisance being carried through these narrow vales."[66]

Wordsworth's shifting thoughts on railways and his argument against the railway in the sonnet and the accompanying letter highlights an entanglement of social, cultural, technological and political discourses in a time of rapid and fundamental changes, drawing attention to the Romantics' involvement in the pressing questions of the early anthropocene. In a sense, the texts by Wordsworth, Baillie and Thoreau I discuss in this chapter can be read as indicators of the ways in which the Romantics (and Transcendentalists) engaged with what we now think of as climate "tipping points" – or moments of no return – emerging in the wake of the widespread adoption of steam-powered travel. It would, of course, be an oversimplification to draw a straight line from Wordsworth's personal objections to the Kendal and Windermere railway to twenty-first-century discussions of climate crisis. What I think is worth noticing here, however, is the way that significant revolutions in travel such as the invention of steam-powered trains and ships as well as other forms of motorized

---

62 Wordsworth and Wordsworth, *The Letters of William and Dorothy Wordsworth – The Later Years, Part I*, 299. The year 1825 saw the opening of the world's first modern railway, the Stockton and Darlington Railway in England, so Wordsworth can perhaps be forgiven for not foreseeing the development this form of travel would take over the next two decades.

63 There is no mention in the letter of how Wordsworth's daughter Dora traveled the 50 miles from the Lake District to Preston, but also no argument against using the railway from there.

64 Wordsworth and Wordsworth, *The Letters of William and Dorothy Wordsworth – The Later Years, Part II*, 21.

65 Hsu, "The Romance of Transportation in Wordsworth, Emerson, De Quincey, and Dickinson," 47.

66 Wordsworth and Wordsworth, *The Letters of William and Dorothy Wordsworth – The Later Years, Part II*, 617.

travel have an impact on perceptions of time and space that move far beyond the possibility of simply getting from point A to point B at a slightly faster speed.

As Tsing remarks in "The Art of Noticing", there is often a difference between the "story we know" and the "story we need to know". She discusses this difference in the context of the Oregon forests, whose logging was made possible by the installment of a railway line at the beginning of the twentieth century – arguing that

> [i]ndustrial transformation turned out to be a bubble of promise followed by lost livelihoods and damaged landscapes. And yet, such documents are not enough. If we end the story with decay, we abandon all hope – or turn our attention to other sites of promise and ruin, promise and ruin.[67]

Tsing's project in *The Mushroom at the End of the World*, which I will discuss further in the chapter on "Mushrooms" is to notice the potential of "damaged landscapes",[68] the developments that can, as the Matsutake mushroom she follows throughout the text, emerge from ruin. While Tsing's perspective regards these emergences and entanglements from the perspective of the current moment of climate crisis, it is also worthwhile tracing the roots of the thought processes that led to the damage in the first place, that is the narratives of progress and ruin, and to question, as Tsing does, their proclaimed emergency and universality.

In the context of Tsing's argument, Wordsworth's sonnet emerges as a narrative of ruin,[69] as it is focused on a specific, disappearing experience of the landscape, clinging to the hope that the landscape will, somehow, preserve itself. His resistance to the progress narrative of the railway is, in this sense, not so much revolutionary as conservationist, and hinges on the hope that either "nature" will protect itself, or humanity might see reason and accept his individual position as an aesthetic necessity. Day, drawing on Robert Mayo, argues that the assumption of a "contrast between the peace and fulness of nature and the evil generated by and between human beings" could be seed as "the fulfillment of an already stale convention" by the time Wordsworth drew on it in the railway sonnet.[70] An "ecological perspective"[71] on Wordsworth's work, even one drawing on Bate's suggestion that we read Wordsworth in the light of his awareness of the ecosystem he inhabited and engaged with, must, consequently, also grapple with the incongruences of his evocation of a "beautiful romance of nature" in the name of universalizing his own perspective. In the next part of this chapter, I will read Joanna Baillie's poem "Address to a Steam-Vessel" as a more ambiguous position towards the emergence of steam-

67 Tsing, *The Mushroom at the End of the World*, 18.

68 Tsing, 18.

69 see Tsing, 19.

70 Day, *Romanticism*, 35.

71 Day, 176.

powered travel that oscillates between gratefulness towards the more democratic access to an aesthetic experience of the landscape it provides and sadness at the loss of the sailing ships it replaced. I also argue that Baillie's poem addresses these questions from a complex positionality that challenges Wordsworth's universalizing perspective.

Joanna Baillie: "Address to a Steam-Vessel"

They, who on higher objects scarce have counted
Than church's spire with gilded vane surmounted,
May view, within their near, distinctive ken,
The rocky summits of the lofty Ben;
Or see his purpled shoulders darkly lower
Through the dim drapery of a summer shower.[72]

Joanna Baillie's poem "Address to a Steam-Vessel" – first published in 1823 in the very generically named volume *A Collection of Poems*[73] – imagines steam-powered travel as a democratic experience providing a wider group of travelers with access to the landscape. In this, her understanding of the potential of steam-powered trains and ships runs directly counter to the sense of intrusion conveyed in Wordsworth's letter and sonnet, which highlight the differences in perception of the landscape experienced by those who move through these spaces on foot as opposed to travelers who arrived by train. Both poems describe an aesthetic experience in the third person, but where Wordsworth's perspective is focused on an underlying alliance between the figure of "the pausing traveller"[74] as a lone observer and a vision of "nature" that is both personified and imbued with the agency to act on behalf of a Romantic aesthetic experience, Baillie's poem introduces the aesthetic experiences available to an undefined "them", juxtaposing the views available to "them" in the city to the beauty of "nature". Its main argument hinges on the idea that the steam-vessel gives a much

---

72 Baillie, "Address to a Steam-Vessel," 41.

73 Baillie is better known as a playwright – indeed Mellor argues that she was "the most influential playwright of the day" (Mellor, *Romanticism and Gender*, 7.) – but she also wrote poetry and edited *A Collection of Poems* (see Baillie, *A Collection of Poems*.) to which she also contributed. Curran opens his seminal discussion of women writers of the Romantic period on a contemporary eulogy of Baillie's work in *Blackwood's* (published in Baillie's native Scotland, but nevertheless a highly influential and widely read source at the time) which suggests that she was, indeed celebrated as a poetic genius in her day (see Curran, "Romantic Poetry: The I Altered," 185.).

74 Wordsworth, *Kendal and Windermere Railway – Two Letters Reprinted from the Morning Post*, 4.

more varied group of travelers – introduced as those "in populous city pent"[75] – access to the landscape.

Democratizing an Aesthetic Experience of Landscape

One of the most striking differences to Wordsworth's sonnet is the construction of positionality in "Address to a Steam-Vessel". The speaker does not include herself among the travelers, instead describing from the outside, how "they" scatter the deck of the steamer with "[b]askets, and cloaks, and shawls of scarlet dye" while "dogs and children through the crowds are straying, / And, on his bench apart, the fiddler playing" turning the whole experience into a "floating fair".[76] The speaker groups everyone but herself into the "motley throng", describing individuals, such as the "matron dames" repairing to their "tressel'd seats", but does not identify with anyone in particular.[77] The effect is not so much a universalization of a personal experience, as in Wordsworth's sonnet, but rather an impression of a diverse group whose experiences are validated, even if they are not fully shared in by the speaker. Despite this somewhat ambiguous introduction of its passengers, the poem's fourth stanza discusses the steam-vessel's merit of opening "nature" up to them, allowing them the unfamiliar aesthetic experience of their natural surroundings:

> Thou hast to those "in populous city pent"
> Glimpses of wild and beauteous nature lent;
> A bright remembrance ne'er to be destroyed,
> Which proves to them a treasure, long enjoyed,
> And for this scope to beings erst confin'd,
> I fain would hail thee with a grateful mind.[78]

The poem's speaking voice, emerging for the first time in its fourth stanza and in the context of praise of the steam-vessel for the access to the landscape it has granted *others*, takes a curious position between personal gratitude and a more philanthropic universalization of gratefulness. While the line "I fain would hail thee with a grateful

75 Baillie, "Address to a Steam-Vessel," 40. By framing the city dwellers in the context of a line from Milton's *Paradise Lost*, Baillie constructs them as part of the aesthetic experience described in the poem (and not the uneducated outsiders that Wordsworth perceives them to be in the letter discussed above).

76 Baillie, 39. I have chosen to read the speaker as female here – not because of the poet's gender presentation, but because the speaker adopts, as I argue throughout this chapter, a complicated subject position that I read as an explicit strategy in women's poetics of the Romantic period.

77 Baillie, 39.

78 Baillie, 40.

mind" suggests a personal involvement of the speaker's emotions, this seems to run counter to the speaker's disappearance behind more general arguments in the rest of the poem. There are two other instances in the poem, where the speaker emerges to express a personal opinion: a "we" in line 91, which emerges as part of a counterargument that foregrounds the aesthetic advantages of sailing vessels over the steamer's qualities of technological innovation, reliability and the democratization of travel; and the "I" expressing a mixed emotional reaction at the end of the poem. These glimpses of an overt speaker are presented as personal opinion and emotional reaction and set in the context of a more general argument that is curiously detached from the emotions of the speaker. Where Wordsworth's sonnet universalizes a very specific experience while at the same time arguing against granting access to this experience for too many others, Baillie's poem celebrates the fact that an aesthetic experience is now available to a larger group of people, presenting the speaker's own mixed feelings about this access as a complex set of mixed emotions that does not countermand the general argument of the poem.

Baillie's argument about the democratization of travel runs directly against Wordsworth's quoted above, suggesting that "wild and beauteous nature" can be glimpsed even by those "erst confin'd" in the city and can serve as "a treasure, long enjoyed" which can also be revisited as "bright remembrances ne'er to be destroyed". For Baillie, neither the aesthetic experience, nor the ability to recollect it later is based on any special predisposition or education (or poetic genius), but open to a "motley" group of city dwellers, who may profit from it even if they are not endowed with any specific aesthetic sensibilities. The relatively simple form of the poem, which consists of 130 lines of iambic pentameters rhyming in pairs with very few minor deviations from this pattern, underlines this understanding of aesthetic experience as open and communal. Wordsworth's letter, by comparison, offers a long discussion of "susceptible taste" and how it might be curated even in "uneducated persons" through "little excursions with their wives and children among neighbouring fields, whither the whole of each family might stroll". The "artisans and labourers, and the humbler classes of shopkeepers" that he instrumentalizes in his argument against the railway, "can be trained" to profit from an experience of "nature" "where she is the most distinguished by the majesty and sublimity of her forms", but

> [s]urely that good is not to be obtained by transferring at once uneducated persons in large bodies to particular spots, where the combinations of natural objects are such as would afford the greatest pleasure to those who have been in the habit of observing and studying the peculiar character of such scenes.[79]

79 Wordsworth, *Kendal and Windermere Railway – Two Letters Reprinted from the Morning Post*, 9.

It is, of course, possible to read this as an argument for the aesthetic education of the masses that might be further understood as the beginning of a localized ecocritical project in the sense suggested by Bate (which I have discussed above), but Wordsworth uses his suggestion to educate those not already endowed with aesthetic sensibilities as a distancing mechanism and counterargument to the building of a railway for better access, demanding that the masses be educated first, and definitely far away from his rural retirement. Where Wordsworth, thus, sees the railway as an intrusion – not just based on its own destructiveness, but also as a means of conveying "uneducated persons in large bodies" into a landscape that might be destroyed by them or their limited tastes, Baillie's poem expresses gratitude to the steam-vessel for broadening the "scope" of the aesthetic experience of "nature" to those "erst confin'd" in the city.

Both Wordsworth's and Baillie's arguments encompass large groups of people, but where Wordsworth's point is exclusionary on the basis of educated taste, Baillie's is inclusionary, suggesting that the experience of "nature" itself would be beneficial to the travelers, regardless of their background and the development of their taste. Moreover, while both texts foreground a personal experience, Wordsworth's sonnet attempts a universalization of his argument (as I have shown above), while Baillie's poem suggests the speaker's individual gratitude to the steam-vessel for providing this opportunity: "I fain would hail thee with a grateful mind."[80] This personal expression of gratitude is indicative of the speaker's positionality in Baillie's poem, which presents the speaker's point of view as individual, not universal (unlike Wordsworth's covert speaker). The use of "I gaze" in the last lines of the poem underlines this construction, as it frames the speaker's emotions in the immediate context of looking at the steam-vessel.

The Aesthetics of Steam-Powered Travel

It would, nevertheless, be an oversimplification to read Baillie's poem solely in its capacity to function as a counterargument to Wordsworth's sonnet. On the one hand, Baillie's poem predates Wordsworth's by more than two decades – a time period, moreover, which saw an exceptionally rapid growth in steam-powered factories and travel. On the other hand, the poem does not solely function as a celebration of the democratic travel opportunities presented by the steam-vessel, but instead hinges on a more complicated, ambiguous argument, which oscillates between the main poles of aesthetics and ethics: while the speaker expresses gratitude for the steam-vessel offering the opportunity of an aesthetic experience of the landscape to the "motley throng" of "passengers of every sort" introduced in the first few lines, the last three stanzas express the speaker's elegiac memory of "light skiffs" as "dearer

80 Baillie, "Address to a Steam-Vessel," 40.

to fancy, to the eye more fair", compared to which the steam-vessel emerges as "a daily lab'rer, a mechanic swart, / In working weeds array'd of homely grey."[81] The last stanza, in particular, draws a concluding connection between the steam-vessel as a "laborer" and "modern men with all their alter'd ways", suggesting not only an understanding of this form of transport as a harbinger of fundamental change, but also expressing a set of mixed emotional reactions to its power in the last two lines of the poem: "Fitful I gaze with adverse humours teased, / Half sad, half proud, half angry, and half pleased."[82] Apart from the mathematical impossibility of this sentence (likely a result of the meter, into which "one quarter" would not have fit), the "adverse" emotions brought forth by the speaker's affective engagement with the steam-vessel are not only contradictory, but also highlight the difficulty of an unencumbered aesthetic experience in the age of steam-powered ships and "modern men".[83] The poem, thus, contrasts the speaker's gratefulness for the possibility of an open and communal aesthetic experience of "nature" with a sadness and anger over the loss of a more personal one associated with watching sailing vessels that has now been eradicated by the technological innovation of steam-powered travel.

While both Wordsworth's and Baillie's poems, thus, convey a sense of irrevocable change, they do so from different positions and with different degrees of complexity. Where Wordsworth's sonnet categorically rejects the intrusion of the railway into a specific "nook of English ground" as "rash assault", Baillie's poem weighs advantages and disadvantages of steamboats, juxtaposing personal sentiments with a discussion of the accessibility of aesthetic experience. The two longer middle stanzas of the poem (stanza 3 of 24 lines and stanza 4 of 46 lines), are dedicated to steam power and to the way it facilitates an experience of "nature" to "eyes that have no fairer outline seen / Than chimney'd walls with slated roofs between, / Which hard and harshly edge the smoky sky."[84] As the remainder of the poem's seven stanzas vary in length between 8 and 16 lines, with stanza 2 and 7 creating a sense of stability at

81 Baillie, 41.

82 Baillie, 41.

83 Baillie, 41. As Madelon Hoedt summarizes, Baillie was no stranger to new technologies. As a prolific playwright she was celebrated for her Gothic plays at a time, when "theatres gr[e]w in both size and number, and as stagecraft [became] more technically sophisticated, with the Gothic as an early adopter of the new possibilities." (Hoedt, "Gothic Drama and the Uncanny Stage," 455.)

84 Baillie, "Address to a Steam-Vessel," 41. The landscape described in the poem is close to Glasgow, one of the centers of industrialization in Scotland from the 1820s onwards. While the city-dwellers that Baillie describes have never had the opportunity to leave the city and enjoy its natural surroundings, the industrialization process in the 19th century also relied on immigrants into the city from the Scottish Highlands (and later also from Ireland), although factory workers are, perhaps, not the main class of travelers that Baillie imagines.

12 lines each, the sheer length of stanza 4 lends a certain gravity to its arguments, presenting them as central to the poem.

Stanza 4 consists of an extended juxtaposition of city and "nature" in the aesthetic experience of the city dwellers, which is, as the poem argues enriched and widened in "scope" by means of the new-found access to the countryside that they can now see on board the steam-vessel. "Nature" is introduced as "wild and beauteous" in line 46, followed by a discussion of more specific landmarks and their features. While the alliteration of "Lochlomond's fair and fairy isles"[85] conveys a hint of unreal beauty of the scenery, the reference to the Loch itself positions the steamer, and with it the perspective of the poem at the southwestern end of the Scottish Highlands. The other landmarks – "lofty Ben"[86] and the Clyde mingling "his waters with the briny tide" of the Clyde Estuary – support this position. The local references to "the lesser Cumra" and "Aron" present something of a puzzle, but likely refer to Wee Cumbra and Arran – two islands in the Firth of Clyde. Loch Lomond and the Clyde are connected by the river Leven, which is no longer considered shippable, but could be traversed by steamers at the beginning of the nineteenth century.[87]

The stanza ends with a view of "fair Ireland's coast", which envisions a sea voyage slightly beyond the Firth of Clyde for the passengers to be able to distinguish the "marshalled pillars" of the Giant's Causeway across the Irish Sea. The poem itself suggests that the latter would have been possible, as it describes the steamer as "[r]over at will on river, lake, and sea".[88] While this view of Ireland's coast is situated in the present of the poem through the use of "now" in line 85, the last lines of stanza 4 introduce a more sinister image in the comparison of the pillars to the ranks of "[p]haraoh's army, on the Red-sea shore, / Which deep and deeper went to rise no more."[89] While the main function of this comparison seems to be a visualization of the natural spectacle of the Giant's Causeway, the biblical allusion also vastly extends the time frame of the poem into the past and evokes the image of drowning in the

85 Baillie, 41.

86 Presumably this refers to Ben Lomond, not the more famous Ben Nevis. "Ben" in line 62 rhymes with "ken" in line 61, introducing a faint hint of local color, although Baillie, who was born in Glasgow, was Scottish and a number of her poems are written in Scots or contain a more substantial amount of Scots vocabulary.

87 According to A. Graham Lappin's description of the local history of steamers on Loch Lomond, the steam-vessel described by Baillie in 1823 was most likely the Marion, built by the engineer David Napier in 1816, which first traveled up the Leven in 1818 to become a popular tourist attraction on Loch Lomond. (Lappin, "The Loch Lomond Steamers."). Baillie's poem follows the reverse route, taking the steam-vessel from the Loch to the sea, most likely to facilitate the move to sea-faring sailing ships in stanza 5 and 6. While Baillie spent most of her professional life in Hampstead, she frequently visited Scotland throughout her life. It is highly likely that she saw the steamer on Loch Lomond firsthand on one of these visits.

88 Baillie, "Address to a Steam-Vessel," 42.

89 Baillie, 42.

sea – thus indirectly envisioning a ghastly end to the voyage described in stanza 4, before the poem moves on to a more general discussion of steam-vessels in stanza 5: "Yet ne'ertheless, whate'er we owe to thee, / Rover at will on river, lake, and sea". The use of "nevertheless" somewhat deflates the arguments made for the usefulness of the steamer to the city population, introducing the stanza's shift to the superior aesthetic appeal of sailing boats. The poem is not concerned with the distinction between the purpose of people's journeys – "as profit's bait or pleasure's lure engage" them – but instead moves the gaze away from what can be seen from the steamer to how the steamer compares aesthetically to the more elegant sailing ships.[90]

The Steamer Marion (from a handbill)

Lappin / valeofleven.org.uk

Aesthetic Experience Beyond the Autonomous Self

Similar to the "bright Scene from Orresthead" in Wordsworth's sonnet[91], the scenery described in Baillie's poem is also localized and, at least to an extent personified, although never addressed directly, as the addressee of the poem is, as the title suggests, the steam-vessel itself. Instead, the description of the landscape includes "bushy promontories fondly peep[ing], / At their own beauty in the nether deep" of the mirroring Loch and "birch and berried row'n that lave / Their vagrant branches in the glassy wave" in lines 55 to 58.[92] This is, as in Wordsworth's sonnet, a feminized "nature", but one that is engaged in the admiration of its own beauty and not roused

90 Baillie, 42.

91 Wordsworth, *Kendal and Windermere Railway – Two Letters Reprinted from the Morning Post*, 4.

92 Baillie, "Address to a Steam-Vessel," 41.

to fend off "rash assault".[93] While these views of "nature" are introduced as part of the passenger's experience as "seated at their ease [they] glide along Lochlomond", the speaker's words direct the gaze towards the mirroring surface of the water reflecting the surroundings.[94] The poem's focus on this area between land and Loch and later between land and sea in the "interlacing slopes that gently merge / Into the pearly mist of ocean's verge" in line 81 to 82 creates a sense of temporal and spatial liminality that the steamer traverses and allows its passengers to experience.[95] The sense of traveling in the liminal spaces between water and land also presents a parallel to the temporal transition from historical sailing vessel to modern steamer, underlining the poem's sense of a significant shift in the world that it attempts to frame emotionally as well as aesthetically.

As in Wordsworth's sonnet and letter, the introduction of steam-powered travel signals the end of an era, and this imagined or postulated transition is also negotiated via an aesthetics that draws on nature imagery in Baillie's text. The personification of promontories peeping fondly, trees dipping their branches in the Loch and "the lofty Ben" darkly lowering his "purpled shoulders [...] / Through the dim drapery of a summer shower" creates a sense of benevolence and connection between the experience of the city dwellers and the landscape, which, to an extent, also includes the interaction with unfamiliar (to the city dwellers) animals – "the sea-mew" and "[k]ittywakes" (seagulls) referred to in line 71 and 74 respectively.[96] The concept of "nature" that emerges from the benevolent, personified elements of the landscape and the experience of "gliding" through it on board the steamer is focused on these picturesque elements rather than a sense of connection with the non-human natural world – mainly because the speaker does not describe a personal experience of "nature", but rather the one available to the travelers on the steamer who are also observed from the outside, allowing the speaker to rationalize an observed aesthetic experience.

In Baillie's poem, city dwellers exchange the limited views of the city with the much more picturesque views of the landscape, however, compared to the immersion and deep emotional involvement in "nature" evoked by Wordsworth, this experience is predominantly visual and focused on the picturesque rather than the sublime. The adjectives referring to the city are, by contrast, negatively connoted, beginning with the relatively neutral "populous" in line 45, followed by the alliterating "hard and harshly" as well as "smokey" (referring to the sky) in line 77, and culminating in the "work of sordid skill, / The storied structure of the cotton mill"

93 Wordsworth, *Kendal and Windermere Railway – Two Letters Reprinted from the Morning Post*, 4.

94 Baillie, "Address to a Steam-Vessel," 41.

95 Baillie, 42.

96 Baillie, 41.

in line 83–4.[97] The curious parallel construction of "sordid"[98] and "storied" suggests that the poem's main criticism of the cotton mill is an architectural, and therefore aesthetic one, as the text also describes steam power in general in positive terms. While the stanza dedicates the most space to describing the beautiful elements of the landscape, descriptions of life in the city are interspersed throughout, associating it with scarcity and the deprivation of aesthetic experience. Most of these focus on architectural elements ("church's spire" in line 60 and the abovementioned "chimney'd walls" and "slated roofs"), which unfavorably contrast with the splendor of the landscape. Even the few glimpses of "nature" available in the city – the "suburb orchards" – are "choked with colworts green", supporting the impression of being "confined" expressed in line 49 and the sense that the city offers "nought of verdant freshness" in line 51.[99] Deprivation also emerges in a socioeconomic sense, clarifying the poem's understanding of the "motley throng" of passengers to include the "city cobbler", "who hath but warr'd with thieving cat, / Or from his cupboard chaced a hungry rat" and is now given the opportunity to engage with the wildlife in the form of "the wild sea-mew in its mid-flight".[100] This is not a majestic or sublime understanding of "nature" and the cobbler's experience is not that of an autonomous self fusing with "nature" in a process of sublimation. Instead, the poem presents the idea of experiencing the non-human natural world as a small personal pleasure to be drawn on in the future as an enrichment to an otherwise limited life in the city. That this experience is not subjective, but available to many different people underlines the poem's point about steam-powered travel as a communal means of accessibility to the aesthetic experience of the non-human natural world.

The predominant mode of experience in the juxtaposition of city and country is visual, with a few appearances of sounds in the form of bird cries. While the references to the city do not include any noises, the first description of the steamer in the second stanza introduces not only its "merge'd wheel[s]" but also the "noisy fray" of the "white-churned waters", suggesting that a comparison between city and a countryside traversed by steamer or train would not come out in favor of the country-

---

97 Baillie, 41.

98 See also my discussion of "sordid" in the Introduction.

99 Baillie, "Address to a Steam-Vessel," 41. This is not the world of suburban gardening celebrated by Elizabeth Kent's *Flora Domestica*, which I discussed in the chapter on "Anthropocene", – rather, in the poem's contrasting view of city and landscape, the "suburb orchards" are also presented as a limited access to the "verdant freshness" of the open landscape that is only accessible through the means of steam travel.

100 Baillie, 41. The fact that the rat is described as "hungry" suggest that the cobbler's cupboard is not well stocked, possibly because of a loss of trade resulting from the incipient industrialization of shoemaking. In any case, the poem introduces the cobbler as a member of the working class which is, throughout the poem, aligned with the vessel itself as a "daily lab'rer" wearing "working weeds" (Baillie, 43.).

side.[101] The comparison of steam- and sailing vessels is also predominantly visual, but suggests a perception of the "light skiffs, that to the breezy air, / Unfurl their swelling sails of snowy hue / Upon the moving lap of ocean blue" as altogether lighter and, by inference, less noisy. This is supported by the metaphors "gentle nymph or lady gay" in line 124 which are used in direct opposition to the "daily lab'rer" and "mechanic swart" to distinguish sailing ships from steam-vessel.[102]

This juxtaposition is further underlined by a temporal distinction in which the "modern men" and "altered ways" are set in the context of a historical dimension that reaches back via Columbus and Magellan to a time "[e]re yet the reign of letter'd lore began", establishing the sailing ships within a long tradition of adventurous travel. While the last stanzas of the poem pit the steam-vessel as a "[r]over at will on river, lake, and sea" against the sailing ships whose need of "manly skill" to "brave [...] the winter blast" lent them a sense of "high adventure", the third stanza celebrates the steam-vessel's "independent pride: / No leave ask'st thou of either wind or tide. / To whate'er point the breeze, inconstant, veer, / Still doth thy careless helmsman onward steer."[103] The juxtaposition of steam- with sailing boats also hints at the underlying transfer of labor from the sailors' "manly skill" to the steam-vessel itself as a "daily lab'rer" loosely controlled by a "careless helmsman".[104] I read "careless" as a positive expression of not having to care about the elements here, rather than an expression of mistrust towards the person at the wheel. The shift in the crew's skills and responsibilities from sailing- to steam-vessel also has wider ramifications: as steam power replaces manpower, the deck of the ship becomes a space of leisure, not work, as it is opened up to travelers and their aesthetic experience of the landscape. The technical innovation of steam power, thus, also creates a shift in gendered accessibility to previously male-connoted spaces.

Gendering Steam Power

The closest Baillie's poem comes to commenting on the ways in which gender expectations affected women is, perhaps, in the "matron dames [who] to tressl'd seats apart repair" (in line 7).[105] Occurring relatively early in the poem, the comment quite literally sets the female travelers of an elevated class apart from the motley "crowd" described in the first stanza, positioning them at an observing distance from and availing themselves of a more luxurious type of seating than the other passengers, who sit on benches. This seating arrangement for a specific type of female traveler

101 Baillie, "Address to a Steam-Vessel," 39–40.

102 Baillie, 43.

103 Baillie, 40–43.

104 Baillie, 40.

105 Baillie, 39.

("matron dames") runs counter to the democratic idea of the "motley throng" whose "scatter'd […] / Baskets, and cloaks, and shawls of scarlet dye" clutter the deck in a material mixture that clearly does not separate by gender.[106] The differentiation in seating, thus, introduces class as an additional category of separation, as working-class women were neither held to the same standards of propriety nor afforded the same privileges of protection as "ladies", complicating the cultural construction of gender roles and stereotypes. I will further address the ramifications of steam-powered travel for women in my conclusion. Here, I would like to draw further attention to the way Baillie's poem addresses gender as an aesthetic category that emerges as more complex than the dichotomy of a male-connoted observing subject and a female-connoted observed natural object in Wordsworth's poem.

The necessity of explaining steam power to laymen and, perhaps more importantly, laywomen in the poem hints at the newness of steam-powered travel. Technical innovation – traditionally male connoted in the shape of "that philosophical sage, / Watt, who in heraldry of science ranks" – is only introduced towards the end of the text, while the third stanza still compares this power unchecked by the elements to magic: "As if the stroke of some magician's wand / Had lent thee power the ocean to command."[107] This magical power is introduced as steam in the next lines, by drawing on the imagery of a conventionally feminized domestic space, rather than the conventionally masculinized inventor's workshop or factory:

> What is this power which thus within thee lurks,
> And, all unseen, like a mask'd giant works?
> Ev'n that which gentle dames, at morning tea,
> From silver urn ascending, daily see
> With tressy wreathings playing in the air,
> Like the loos'd ringlets of a lady's hair;
> Or rising from the enamell'd cup beneath,
> With the soft fragrance of an infant's breath:[108]

Even without the direct reference to "gentle dames", there is a sense of domestication and feminization in this description that draws on the familiarity of "morning tea" and the comparison to "a lady's hair", suggesting an attempt at introducing an – at this point – fairly unfamiliar form of power as harmless and quotidian. Esther Schor describes Baillie's "genius for demystification" in her discussion of another

106 Baillie, 39.

107 Baillie, 40–42. This reading of steam power as magical was not uncommon, as Davies shows, quoting, for example, Walter Scott, who eulogized Watt as a "magician, whose cloudy machinery has produced a change on the world" (Davies, "Introduction: Romantic Studies and the 'Shorter Industrial Revolution,'" 190.).

108 Baillie, "Address to a Steam-Vessel," 40.

poem on tea – "Lines to a teapot".[109] The audience to be convinced by these comparisons is broadened to include the working classes, as the poem moves from comparing steam power to the innocuous and quintessentially British custom of tea towards that of a homely lunch, bridging the gap between different classes in the process:

> That which within the peasant's humble cot
> Comes from th' uncovere'd mouth of sav'ry pot,
> As his kind mate prepares his noonday fare,
> Which cur, and cat, and rosy urchins share:[110]

Here, as in the earlier comparison, children ("infant's breath" and "rosy urchin") are introduced to suggest not only domestic bliss, but also the quotidian harmlessness of steam power. A temporal component ("daily", "noonday") suggests that steam is already part of everyday life and, therefore, not something to be afraid of. Its "naturalness" is introduced in the following lines, adding a further component of wonder and awe to this general advertisement for steam power:

> That which, all silver'd with the moon's pale beam,
> Precedes the mighty Geyser's up-cast stream,
> What time, with bellowing din exploded forth,
> It decks the midnight frozen north,
> Whilst travellers from their skin-spread couches rise
> To gaze upon the sight with wond'ring eyes.[111]

While this sudden introduction of the "frozen north" may seem like a curious comparison in a poem about a steamer in Scotland, its position in the poem suggests a specific function, as it occurs between the general introduction of steam power via domestic imagery and the description of how the steam-vessel facilitates travel and, thus, the exploration of otherwise inaccessible landscapes to a population of city dwellers. The geyser is, thus, explicitly introduced as a touristic experience for "travellers [who] gaze upon the sight with wond'ring eyes". The structure of stanza 3 constructs a curious congruence between the images of domestic and natural sources of steam by connecting them with the parallelism of "that which", a structure that is used in answer to the question posed at the beginning of the stanza: "What is this power [...]?" The use of sublime imagery to describe the natural power of steam at the end of stanza 3, thus, serves to create both a passage into the descriptions of landscape in stanza 4 and a standard of sublime wonder from which the poem then

109 Schor, "'Stirring Shades': The Romantic Ode and Its Afterlives," 117.

110 Baillie, "Address to a Steam-Vessel," 40.

111 Baillie, 40.

deviates in order to describe a more benevolent/less spectacular domestic "nature" made accessible by the steamer.[112]

While the poem presents a general tendency to end stanzas with a dramatic image, from the "floating fair", via the spectacle of the geyser observed by travelers "wond'ring eyes" to the sunken army, "ris[ing] no more", these are mainly used as visualizations of previous descriptions that are connected by inference rather than direct observations of the steamer or its route. Stanzas 5 and 6, which introduce sailing ships as both aesthetically pleasing and historically relevant, are no exception to this rule, the first one ending on a simile comparing the movement of the ships ("they change, and veer, and turn") to "living things", while the second dives into the history of "hardy seamen", suggesting it stretches back to a time "[e]re yet the reign of letter'd lore began."[113] The function of these poignant last lines seems to be the creation of suspense, which is, however, seldom picked up by the beginning of the following stanza. While stanza 1 and 2 serve as a general introduction, stanza 3 and 4 directly address the steam-vessel as "thou", followed by phrases introducing counterarguments in stanza 5 and 7: "Yet ne'ertheless" and "[i]n very truth". These support the general structure of the poem which moves from praising the steam-vessel as a means of accessible travel to the comparison of steam and sailing ships, ending on the ambiguous feelings of the speaker in the last stanza – "[h]alf sad, half proud, half angry, and half pleased."[114] While pleasure and pride emerge in the first part of the poem with its focus on the steamer and its possibilities, the second part is ruled by sadness over the end of the long history of sailing ships that the steamer heralds. Anger is less prevalent in the poem itself, but its expression in the last line links Baillie's text to Wordsworth's angry letters and sonnet. It is interesting that, while neither poem expresses quite as dire a sense of the end of the world as Morton's retrospective discussion of the steam engine as the driving force of the anthropocene suggests, both contemporary texts, nevertheless, convey a sense of experiencing a period of fundamental change associated with the steam engine and its impact on how travelers were able to engage with the non-human natural world. What further unites the two poems is their understanding of steam-powered travel as a way of accessing a specific space (even if they express very different opinions on access),

---

112 Eric Gidal quotes an article in *Blackwood's Magazine*, which celebrated steam power as "a new power over nature" and, by extension "a new source of wealth for the nation" (Gidal, "Industrial Transport and Political Economy in Blackwood's Edinburgh Magazine," 291.). Baillie's program in introducing steam power as both natural and domestic seems to differ from this assessment: it is based on familiarity, rather than gaining power and focused on employing this power in the service of an aesthetic experience, not commercial transportation or, indeed, financial gain – although tourism as a mass industry did, eventually, lead to a commodification of landscape and travel as an aesthetic experience.

113 Baillie, "Address to a Steam-Vessel," 42–43.

114 Baillie, 43.

which is discussed from a specific positionality and within a spatiotemporal context that reaches beyond the moment described in the text.

One dimension of accessibility in Baillie's poem is gender, which is also expressed as part of the aesthetic register of the text. As I have shown above, "Address to a Steam-Vessel" directly comments on shifts in the politics of access to the aesthetic experience of landscape through technological innovation, but it also does so in its aesthetic juxtaposition of steam- and sailing vessels: as steam is introduced via a series of feminized and domesticated allusions to gentle ladies and their steaming teapots, sailing vessels are associated with the masculine power of "hardy seamen" descended from "bold sagacious man." While steam itself is, thus, feminized, the steam-vessel is, nevertheless, compared to a "daily lab'rer, a mechanic swart, / In working weeds array'd of homely grey / Opposed to gentle nymph or lady gay," that is, to the feminized sailing vessels.[115] The poem openly introduces this "opposition", drawing attention to a curious construction of balance that reinforces common inscriptions of gender: while the control of sailing ships, traditionally imbued with female attributes (also associated with the sea as a cruel mistress) and often bearing women's names as well as female pronouns underlining this identification, demands masculine strength and skill, the steam-vessel runs on its own (masculinized) labor, as man has been replaced with machinery and a "magical" power which is associated with domestic uses and thus, at least in Baillie's poem, feminized. This comparison is rooted in the relative newness of steam power which still needed to be explained and could, thus, from Baillie's point of view, and presumably also for her readers, be associated with common, steam-producing household items such as tea urns and cooking pots, as well as the natural spectacle of the geyser.

This use of gender as an aesthetic rather than a social category is particularly remarkable in that it manages to remain relatively untethered from the experiences of actual men and women, thus not only highlighting the ability of gendered inscriptions to attach themselves to cultural discourses and objects but also deferring questions of gendered access to aesthetic experience from the discussion of personal experience. There is a distinct difference to Wordsworth's universalizing approach here, which emerges from the combined strategies of a distanced speaker position, a foregrounding of the commonality of experience and the deference of social questions to the aesthetic level of the text. In this, Baillie's approach to steam-powered travel and its innovative potential also emerges as more complex than Wordsworth's drawing attention to the conflicting consequences technological innovations created for women in the early anthropocene.

I will return to the gendered implications of steam-powered travel and accessibility in the last part of this chapter, after discussing Henry David Thoreau's position

115 Baillie, 43.

on the railroad in *Walden* as an example of situated knowledge in an early anthropocene text whose highly influential legacy continues to resonate with current ecocritical approaches to climate crisis. As I will argue, Thoreau's introspective project of self-improvement in "nature" allows him to present an external, situated perspective on the railway not so much as an aesthetic, but as a commercial project with substantial socio-cultural repercussions. Thoreau's detailed perspective also traces changes the railway wrought in the relationship between human society and natural resources as well as shifts in the framework of temporality brought about by "railway time". While I situate Thoreau in the American context, his philosophical engagement with the railway offers important critical context for my argument around the ways in which early anthropocene texts reflect substantial technological and socio-cultural changes.

Henry David Thoreau, *Walden* and the Railway

What's the railroad to me?
I never go to see
Where it ends.
It fills a few hollows,
And makes banks for the swallows,
It sets the sand a-blowing,
And the blackberries a-growing,[116]

Henry David Thoreau's short poem about the railroad, a relatively new invention that had only made its way to his hometown of Concord in the 1830s, seems to express, first and foremost, indifference. The poem opens with a rhetorical question, positing a personal connection between the speaker and the railroad, only to suggest that there is no such connection. The line, however, does not only express indifference towards the railroad, it also suggests the importance of positionality in posing such a question, as this expression of personal indifference implies a certain amount of self-assurance or at least the knowledge that someone might be interested in the speaker's personal opinion about the railroad. The second and third line – "I never go to see / Where it ends" – are equally controversial, starting with the speaker's own position of, assumed, disinterest in following the tracks but also hinting at the possibility of an end that might be worth pursuing for others. This suggestion is contested by the structure of the poem, which consists of rhyming couplets (me/see, hollows/swallows, blowing/growing) with the exception of line 3, which, thus, does

116 Thoreau, *Walden*, 168. While the poem is included in *Walden*, published in 1854, it was more likely written during Thoreau's stay by Walden Pond in 1845–47.

not actually have an ending, because it misses the second part of its rhyming couplet and leaves the conclusion of the poem open. Instead of providing a sense of closure, the second part of the poem sets the railway in connection with the non-human natural world, focusing on the unplanned ramifications of building a railroad instead of its more general progress narrative. In this, the poem does not only inaugurate the general argumentative structure of *Walden*'s approach to the non-human natural world, but it also presents a point of attachment for later environmentalist discourses skeptical of technological innovation that shift the perspective towards its impact on "nature".

While the filling of hollows, the creation of nesting opportunities for swallows and an over-abundance of blackberries may seem like positive effects at first glance, there is no systemic coherence between these unintended and unconnected effects on already existing ecosystems and, while they might have seemed marginal and harmless enough to the reader in Thoreau's time, the blowing away of sand rings alarm bells for more ecologically attuned readers today. Even if Thoreau and his contemporaries were not fully aware of the effects of overlogging associated with the construction and use of the railway and did not consider the blown-away sand in the context of desertification processes, the image with its associations of needless fuss and barrenness does not create a positive effect. Instead, the second part of the poem supports the suggestion of indifference introduced in the first line with phrases such as "a few", which highlight the general unimportance of the effort.[117] The use of double and triple rhymes in lines 4 to 7 conveys a sense of structural balance and simplicity which is undermined by the lack of a rhyming counterpart to the word "ends" in line 3, stressing its importance as well as the openness of the line and the whole poem. As the speaker "never go[es] to see / Where it ends", if and how it ends remains unclear. The poem itself does, in fact, not end, but is broken off mid-sentence, followed by a further discussion of the arriving train as a nuisance: "but I cross it like a cart-path in the woods. I will not have my eyes put out and my ears spoiled by its smoke and steam and hissing."[118] The concept of "crossing" carries an ambiguous meaning in this context, as it does not only suggest a spatial "crossing" of the tracks, but also a mental "crossing" or rejection of the concept of the railway both in Thoreau's thoughts and in his writing, which also switches genre from poetry to autobiographical prose and, thus, from an attempt at aestheticizing his stance to a personal observation of a specific encounter with the railroad tracks in the woods at this point in time.

117 In other instances in *Walden*, Thoreau laments the immense effort it took workers to build the railroad, as well as the many deaths it cost. His disinterest in the railroad itself is, as I will argue below, also accompanied by regret over the toll in human lives it took to build it.

118 Thoreau, *Walden*, 168.

Progress and Commerce

Thoreau's diary entry of 23 June 1853 takes the ecocritical aspects of his engagement with the railway further, suggesting that by 1853, Thoreau had in fact developed an even more pronounced critical opinion of its uses and ramifications than the poem, likely written in the 1840s, discloses:

> I was just roused from my writing by the engine's whistle, and, looking out, saw shooting through the town two enormous pine sticks stripped of their bark, just from the northwest and going to Portsmouth navy-Yard, they say. Before I could call Sophia, they had got round the curve and only showed their ends on their way to the Deep Cut. Not a tree grows now in Concord to compare with them. They suggest what a country we have got to back us up that way. A hundred years ago or more perchance the wind wafted a little winged seed out of its cone to some favorable spot, and this is the result. In ten minutes they were through the township, and perhaps not half a dozen Concord eyes rested on them during their transit.[119]

The diary entry captures the speed of the railway, but also, to an extent the speed of the logging process that stripped the area of a wealth in trees, even if more and larger pines remain. It is clear that Thoreau's awe at the size of the trees is not entirely free of regret. His thoughts revolve around the dimension of time, juxtaposing the speed of the train transporting the tree and the speed of logging on the one hand with the span of a "hundred years" that it took the "little winged seed" to grow in "some favorable spot."[120] There is a definite sense of longing for the trees of the past contained in the comparison of the "enormous pine sticks" on the train with the trees "grow[ing] now in Concord to compare with them", and this is underlined by the additional speed of the railway which whisks these enormous tree trunks out of sight. Thoreau was clearly aware of the ongoing trend of deforestation,[121] although not, perhaps, its environmental consequences. By 1853 there were few forested areas left in Concord and its environs, although some were regrown later on. At the time Thoreau conducted his experiment of living in the woods near Walden Pond in 1845–47, the Fitchburg railway had already been built and Thoreau habitually crossed the tracks in the woods or heard its whistle, which served as a constant reminder of these ongoing processes as well as their cultural meaning and ramifications, which

---

119 Thoreau, "Journal – June 23," 298.

120 Thoreau wrote an essay about "The Succession of Forest Trees" further exploring the process of deforestation and how logged pine trees were often succeeded by oaks, arguing that seeds transported by the wind were the cause for this effect (see Thoreau, *The Succession of Forest Trees and Wild Apples.*).

121 see Buell, *The Environmental Imagination*, 120.

he discusses in his written account of these years. In *Walden* (1854), Thoreau offers perhaps the most condemnatory view of the railway, but he does so in predominantly sociological not ecological terms:

> Men think that it is essential that the *Nation* have commerce, and export ice, and talk through a telegraph, and ride thirty miles an hour, without a doubt, whether *they* do or not; [...] If we do not get our sleepers, and forge rails, and devote days and nights to the work, but go to tinkering upon our *lives* to improve *them*, who will build railroads? And if railroads are not built, how shall we get to heaven in season? But if we stay at home and mind our business, who will want railroads? We do not ride on the railroad; it rides upon us. Did you ever think what those sleepers are that underlie the railroad? Each one is a man, an Irishman, or a Yankee man. The rails are laid on them, and they are covered with sand, and the cars run smoothly over them. They are sound sleepers, I assure you. And every few years a new lot is laid down and run over; so that, if some have the pleasure of riding on a rail, others have the misfortune to be ridden upon.[122]

Thoreau's argument here is focused on the suggestion of the railway as a project of progress (or "manifest destiny" in the US context), which is directly linked to the Christian ideal of a heavenly reward, which Thoreau mocks here, with the idea of getting to heaven "in season" by building a railway. The passage does not only condemn the exploitative working conditions of railroad companies and their practice of drafting recent immigrants from Ireland who had few other choices of making a livelihood, but also the ideas of progress and acceleration, as well as economic necessity that drove the construction of railroads in the first place.[123] Thoreau understands "commerce"[124] as a concept directly opposing the transcendentalists' Neoplatonic ideals which were set against materialist wealth as a hindrance to understanding the true essence of things and the true calling of mankind.[125] This benevolent anthropocentrism emerges in *Walden*, for instance in Thoreau's juxtaposition of the concepts of the railroad and telegraph as improvements with the project of self-improvement and "deliberate" living forwarded by Thoreau himself. In the passage quoted above, this discrepancy emerges as an argumentative split between what men think what the "*Nation*" needs and what they themselves might need and want.[126]

---

122 Thoreau, *Walden*, 136–37.

123 see Mangrum, "Nature, Necessity, and the Philosophy of Metaphor in *Walden*," 50.

124 I also discuss the resistance to "commerce" in my Introduction in the context of Maria Logan's work.

125 see Day, *Romanticism*, 186ff.; see also Moore, *Ecological Literature and the Critique of Anthropocentrism*, 157.

126 Gidal draws attention to the complex discussions over whether commerce drove the development of transportation or the other way around. (see Gidal, "Industrial Transport and Political

Progress and the Project of Self-Improvement

For Thoreau, any imagined interest in the railway hinges predominantly on wanting to use it to go somewhere, and his phrasing suggests that most people would, as a matter of fact, rather stay at home and "mind [their] own business"[127] than travel anywhere. This argument rests on the cultural assumption that everyone's "business" would, in fact, be conducted within a radius of their own home that could be reached on foot or horseback and that the railway, therefore, would only be used for the purpose of seeing other parts of the country – that is for touristic purposes. In this, Thoreau's argument against the railway functions similar to Wordsworth's cited above, allowing both to present personal/aesthetic arguments against it, rather than economic ones. Tourism or travel for pleasure depends on interest, more than necessity and is, thus, influenced by people's mindset. It is in this sense that "the railroad rides upon us" because it becomes a self-perpetuating idea that is driven forward by people who don't really want or need it, but upon whose backs it then rests. The added pun on "sleepers" juxtaposes its different meanings, including railway carriages, the wooden beams used in the construction of the railroad and the metaphorical use of "sleeping" as purposefully remaining in ignorance, or as Thoreau puts it earlier in *Walden*, "lead[ing] lives of quiet desperation".[128] Thoreau's criticism in this passage does not engage with the railroad alone[129] – although he expresses the futility of building a railroad for people who might just as well "stay at home and mind [their] own business" – but rather with the idea of improvement and progress through the combined power of technological innovation, acceleration and consumption that is pitted against Thoreau's own project of self-improvement through simplification, self-inspection and deceleration. The railroad is, thus, singled out not only as a symbol of unnecessary technological progress, or on the basis of its intrusion into the "retirement" of the countryside, but as an expression of a mindset that Thoreau actively opposes.

Thoreau's view of acceleration is expressed more clearly in another passage, where he comes close to Wordsworth's view of the advantage of traveling on foot,

---

Economy in Blackwood's Edinburgh Magazine," 281–82.). Gidal's article analyses *Blackwood's Magazine* and while the Scottish situation may be different from what Thoreau experienced, Thoreau's construction seems, for want of a better word, simplified and geared towards an understanding of the individual in society as the driving force behind these ideas of progress.

127 Thoreau, *Walden*, 136.

128 Thoreau, 50.

129 Lawrence Buell argues that Thoreau had an "ambivalent response to technology" which can not altogether be read as objection (Buell, *The Environmental Imagination*, 411.), but Bridget Marshall draws attention to Thoreau's awareness of ecological destruction, lamenting "damage wrought by factories on his beloved river" in *A Week on the Concord and Merrimack Rivers* (1849) (Marshall, *Industrial Gothic*, 156.).

but, once again, frames it within a personalized, socio-economic rather than an aesthetic context:

> One says to me, "I wonder that you do not lay up money; you love to travel; you might take the cars and go to Fitchburg to-day and see the country." But I am wiser than that. I have learned that the swiftest traveller is he that goes afoot. I say to my friend, Suppose we try who will get there first. The distance is thirty miles; the fare ninety cents. That is almost a day's wages. I remember when wages were sixty cents a day for laborers on this very road. Well, I start now on foot, and get there before night; I have travelled at that rate by the week together. You will in the mean while have earned your fare, and arrive there some time to-morrow, or possibly this evening, if you are lucky enough to get a job in season. Instead of going to Fitchburg, you will be working here in the greater part of the day. And so, if the railroad reached round the world, I think that I should keep ahead of you, [...] Men have an indistinct notion that if they keep up this activity of joint stocks and spades long enough all will at length ride somewhere, in next to no time, and for nothing; but though the crowd rushes to the depot, and the conductor shouts "All aboard!" when the smoke is blown away and the vapor condensed, it will be perceived that a few are riding, but the rest are run over, – and it will be called, and will be, "A melancholy accident." No doubt they can ride at last who shall have earned their fare, that is, if they survive so long, but they will probably have lost their elasticity and desire to travel by that time.[130]

I have quoted this passage in its entirety to show its awareness of class differences, which are woven into the image of being "run over" by the departing train in spite of its alluring narrative of "rid[ing] somewhere, in next to no time, and for nothing". In difference to Baillie, Thoreau does not, however, address the problem of accessibility, assuming instead, that the unidentified addressee is able to work to gain their train fare at the same rate as Thoreau could walk to Fitchburg on foot (both assumptions that are complicated by questions of gender, class and disability, for instance).

As elsewhere in *Walden*, Thoreau's arguments are linked to broader political, socio-economic and philosophical questions, but rooted in the self-sufficiency at the center of his experiment in living. The text gives a meticulous account of how he balanced his expenses against actual or potential wages for occasional jobs. There is a certain symmetry to the passage based on the equivalence of a day's wages to the cost of one train ticket, but this also depends on the symmetry of time introduced by the assumption that both contestants would start with nothing (not having, in fact, "lay[ed] up money" for traveling) as well as the assumption that neither would have a reason or purpose to go to Fitchburg, other than the love of travel and seeing

---

130 Thoreau, *Walden*, 96–97.

the country, so that it would not actually matter when they got there. This simplification serves the purpose of the tale as an exemplary allegory introduced to highlight the economic ramifications of traveling on the railway. Compared to Joanna Baillie's "Address to a Steam-Vessel", which praises the democratization of travel through the accessibility and affordability of the steamer for many people, Thoreau's argument is instead focused on highlighting the value of one's own command of time over the necessity of making money, because the given inequalities hinted at in the second part of the passage will always lead to casualties and there is no guarantee that anyone will be able to enjoy the fruits of their labor. Instead, Thoreau suggests that one should live fully in the moment – a sentiment encapsulated here in the phrase "the swiftest traveller is he that goes afoot", but which also occurs elsewhere in *Walden*. Steam-powered travel, which for Baillie heralds the accessibility of an aesthetic experience otherwise not available to the public, remains a suspicious innovation for Thoreau who is more interested in its cost on a social and personal level. Moreover, his distrust in the acquisition of new things (he lists clothes and furniture as examples in the chapter "Economy") is here extended to the railroad as a new and costly acquisition for the nation. Richard Ruland and Malcolm Bradbury describe Thoreau as "a symbolist [...] seeing simultaneously the thing before him and the spiritual analogue he held equally real."[131] Considering the railroad in the light of the ideas it stands for, the passage oscillates between allegorical and literal meaning to ultimately present an argument not just against the railroad itself, but against the commodification of travel and, by extension, the idea of capitalism as exploitative and, ultimately, detrimental to people's lives.

Thoreau's expression of anti-capitalist ideas hinges on his distrust in turning what could be obtained and given freely into a commodity: "I have since learned that trade curses every thing it handles".[132] While the idea of "spending of the best part of one's life in earning money in order to enjoy a questionable liberty during the least valuable part of it", thus, comes close to Wordsworth's "plans of retirement sown in youth"[133], but expresses a contrary idea: Thoreau's phrasing suggests that one should rather use one's youth to live in the moment, instead of wasting it in making plans for retirement.[134] Thoreau, in comparison to Wordsworth, also extends this argument across class boundaries to the railway workers, whom he sees as "brothers of mine", whose time and lives are valuable, even if they have been misled by common narratives of progress and the work needed to achieve it:

131 Ruland and Bradbury, *From Puritanism to Postmodernism – A History of American Literature*, 128–29.

132 Thoreau, *Walden*, 113.

133 Wordsworth, *Kendal and Windermere Railway – Two Letters Reprinted from the Morning Post*, 4.

134 This argument runs directly counter to the Puritan (and general Christian) work ethic, which suggests a reward for hardships and labor in the afterlife.

> "What!" exclaim a million Irishmen starting up from all the shanties in the land, "is not this railroad which we have built a good thing?" Yes, I answer, *comparatively* good, that is, you might have done worse; but I wish, as you are brothers of mine, that you could have spent your time better than digging in this dirt.[135]

The phrasing of the passage suggests a heated political debate, drawing perhaps on Thoreau's experience of voicing his arguments in public, but the imagined interjection is also linked to social criticism, suggesting that "a million Irishmen" still live in "shanties" even after they have built a railroad. Thoreau is not just concerned with his own time, but also with the political project of letting others see how they might spend their "time better than digging in this dirt" – or, presumably, doing other kinds of senseless and badly paid labor in the name of progress. In comparison to Wordsworth's clear condemnation of the railway as an intrusion into the "retirement" of the countryside and his own, Thoreau's *Walden* presents a more complex view of the ways in which the acceleration of the railway changes his perception of the landscape surrounding Walden Pond and, perhaps more importantly, his view of its sociocultural impact and the way the railway changed people's perception of and relation to time and space.

Situating Knowledge in the Experiencing Self

Where Wordsworth's sonnet "On the Projected Kendal and Windermere Railway" presents an attempt at universalizing the speaker's (and by extension Wordsworth's) own opinion, Thoreau's opinions in *Walden* are deeply rooted in his own experience, which is then contrasted with a more generalized opinion, indicating his awareness of being at odds with society – or at least the more progress- and commerce-oriented parts of it. Thoreau frames his experiment of living "deliberately" as personal on the first pages by clearly addressing the problem of positionality in his writing:

> In most books, the *I*, or first person is omitted; in this it will be retained; that, in respect to egotism, is the main difference. We commonly do not remember that it is, after all, always the first person that is speaking. I should not talk so much about myself if there were any body else whom I knew as well. Unfortunately, I am confined to this theme by the narrowness of my experience. Moreover, I, on my side, require of every writer, first or last, a simple and sincere account of his own life, and not merely what he has heard of other men's lives; some such account as he would send to his kindred from a distant land; for if he has lived sincerely, it must have been in a distant land to me.[136]

135 Thoreau, *Walden*, 97.

136 Thoreau, 46.

This, once again ironically phrased, expression of positionality is rooted not so much in poetics, but in Thoreau's transcendentalist background, drawing not only on philosophical self-reliance, but also on an understanding of the self as capable of reaching a high degree of self-knowledge through contemplation. What to today's ears may sound like a witty aphorism is, thus, actually an expression of the limitations of thought and experience. While Thoreau's tendency to formulate aphorisms may well account for the continued circulation of quotations from *Walden* (the sentence "I should not talk so much about myself if there were any body else whom I knew as well" certainly qualifies as a meme for the social media generation) his poetics is rooted in a fundamental understanding of the limitations of knowledge as situated in one's own perception and contemplation. As Thoreau's project throughout *Walden* is focused on simplifying life, it is, necessarily his own empirical experience he foregrounds and catalogues. To "require of every writer, first or last, a simple and sincere account of his own life", however, moves beyond this self-reflection into a prescriptive poetics focused on life-writing and requiring simplicity and truth from other writers as well as himself. Ruland and Bradbury summarize the contradictory project set forth in *Walden*: "Thoreau insists that we learn from him but rejects any direct attempt to follow his particular path from cabin to pond; he had cabin and pond, we have *Walden*." In their view "*Walden* is an experience, as well as the description of one."[137] Part of this ambiguity emerges from the multiple roles that Thoreau takes up in the text, oscillating between the person experiencing and the writer analyzing and framing the experience with purpose and aesthetics in mind, as in "*Walden*, simple narrative is persistently lifted and transformed by its context."[138]

The suggestion that if anyone has "lived sincerely, it must have been in a distant land to me" expresses Thoreau's discontent, not only with the way other authors describe their lives, but, ultimately also, how they live them. The passage, thus, ends on a quip expressing discontent with the current sociopolitical situation, but the discussion of self-knowledge is useful in showing the self-reflexivity of Thoreau's position – especially with regard to the common practice of obscuring the speaking "I" in favor of a universalization of individual ideas. Thoreau's overt adoption of an autodiegetic perspective, thus, also includes a meta-diegetic component as it allows him to reflect common writing practices and to deliberately oppose them by presenting his experiment from a personal perspective and as personal experience. This does not preclude a didactic project – after all Thoreau is interested in convincing his readers of his opinions – but it specifically presents his experiences as situated knowledge by discussing the situatedness of the autodiegetic voice and its proposed readership:

137 Ruland and Bradbury, *From Puritanism to Postmodernism – A History of American Literature*, 126.

138 Ruland and Bradbury, 127.

> I would fain say something, not so much concerning the Chinese and Sandwich Islanders as you who read these pages, who are said to live in New England; something about your condition, especially your outward condition or circumstances in this world, in this town, what it is, whether it is necessary that it be as bad as it is, whether it cannot be improved as well as not.[139]

The readers that Thoreau envisions are fellow New Englanders – even people living "in this town" – precisely because of his understanding of personal knowledge as situated within a sociopolitical context and because he sees his personal experiences as most valuable to people in a similar situation. For Thoreau, the personal is political in the sense that his personal stance is expressed with the intent of convincing his fellow citizens and, thus, implementing political change. This construction provides entry points for numerous conservationist, ecocritical and personal readings in the twentieth- and twenty-first century, but it has also created a somewhat more universalizing fandom that reads *Walden* as an ecological or minimalist bible, without addressing the situatedness of Thoreau's project.

Thoreau's gesture of situating his experiences and opinions within a specific context is repeated throughout the text, which often highlights the juxtaposition of what the autodiegetic "I" sees, experiences, thinks and concludes with what others think by using specific keywords. In the passages I have discussed above this includes, for instance, the use of "I, on my side, require" as opposed to "[m]en have an indistinct notion", but there are numerous other examples throughout *Walden*. Where experiences are universalized, it is usually across time, not between individuals or for groups, for instance in phrases such as "I have often found". Where Thoreau uses "we" to include himself, it is in reference to conventions he does not question, such as in the discussion of positionality quoted above.

This construction of a speaking subject that is frequently opposed to general opinions complicates the discussion of concepts such as time, space, life or humanity which are commonly thought of and presented as universal, but which, within the context of Thoreau's situated knowledge, emerge from the discussion of different, usually contradictory positions. In the following I will, therefore, take a closer look at Thoreau's discussion of time as it emerges in the chapter "Sounds" that contains the longest discussion of the railway in *Walden*.

Clashing Temporalities

Following the chapter on "Reading", "Sounds" opens on the juxtaposition of book language and knowledge with "the language which all things and events speak without metaphor, which alone is copious and standard." To perceive it, Thoreau argues,

139 Thoreau, *Walden*, 46.

we need "the discipline of looking always at what is to be seen."[140] The chapter opening, thus, situates the acquisition of knowledge not (or not solely) in reading, but in "the necessity of being forever on the alert",[141] and to immerse himself in his natural surroundings. This emphasis on the importance of experiencing his natural surroundings is rooted in the transcendentalist conception of "nature", "not as a mere commodity" but as an expression of a divine "spiritual world" which "spoke directly to the self, to the individual mind and soul."[142] Instead of "time substracted from my life", this "undisturbed solitude and stillness", thus, allows him to experience a process of growth, that he relates back to the practice of just existing in and contemplating his surroundings.[143] In this process, Thoreau does not only experience time differently, but this different experience of time emerges as one of the central components of his project of "deliberate" living.

While the discussion of time is central to my argument here, space, or rather the situatedness of the experience in spatial terms[144] also plays a role in Thoreau's writing as it bridges his role as a citizen of a specific country with his experience of "nature" in a specific place, thus inscribing himself into a tradition of nature writing on the American continent. In a sense, Thoreau's writing (along with Emerson's) determined what was perceived as the epitome of ecocritical writing in American literature and the basis for later ecocritical ideas. As Antonia Mehnert argues,

> for American Studies scholars, ecocriticism in its early stages was bound to deal with the seminal American nonfiction texts by transcendentalists Ralph Waldo Emerson and Henry David Thoreau [...]. These texts derive from a tradition of thought that focused on a sense of place and an ethics of proximity based on local knowledge as well as an appraisal of the "wild" as the space of some pristine, untouched nature.[145]

This understanding is, as Mehnert argues, tied not only to a specific idea of "nature" but, more importantly, rooted in "a sense of place". Ruland and Bradbury read this situatedness in the context of American literary history, arguing that "the Puritans denied [...] what, two hundred years later, Henry David Thoreau accepted, that American nature was the greatest form of instruction the continent offered, that,

---

140 Thoreau, 156.

141 Thoreau, 156.

142 Ruland and Bradbury, *From Puritanism to Postmodernism – A History of American Literature*, 119.

143 Thoreau, *Walden*, 157.

144 As Haefner argues, for the Romantics "space is not something 'out there' that we passively perceive, nor is it a mental construct that the individual human mind imposes on the external world. Instead, space is controlled by cultural factors and by the behavioral patterns of society." (Haefner, "The Romantic Scene(s) of Writing," 257.)

145 Mehnert, *Climate Change Fictions*, 36.

as Whitman said, the land itself was the greatest poem."[146] For the Puritans, "[t]he [Native American] sense of mystic reverence for the land, of the timeless cycles that separated him from European linear history, had not yet become central to American culture."[147] This curious separation of the Puritan imagination – routed in religious allegory – from the natural surroundings of the writer not only had a lasting impact on American literature, it also led to an understanding of the natural world as ultimately subject to and shaped by external theories and interpretations. The explorers and English settlers were completely oblivious to Native American traditions of husbandry and landscaping, simply because it did not correspond to their ideas of what farming looked like. As Anya Zilberstein argues, they also saw their own impact on a seemingly wild "nature" as beneficial to both landscape and climate.[148] As Ruland and Bradbury argue, this failure to engage fully with the landscape lefts its mark on American literature:

> American nature itself was hard material for the mind to manage. American physical space was vast, its climate varied and often dangerous [...]. It was neither tamed nor enclosed, neither a garden to work on nor distantly sublime and enlarging to the visual and aesthetic imagination. There were those in Europe, like Buffon in France, who saw the American climate as perverse, a danger to man and his development. Others sought to embrace the special and remarkable wonders of American nature, but the American metaphysics that would merge man and terrain was slow to develop.[149]

Ruland and Bradbury's argument sets Thoreau in a context of writing not just "nature" in general, but a specific "nature", which emerges in *Walden* as the intricate description of a specific area and its impact on Thoreau as well as its other inhabitants. The experience described in *Walden* is, thus, rooted in a specific spatiotemporal context and the engagement with this particular area of the non-human natural environment allows Thoreau to "explore the balance between human impermanence, Eurocentric dominance, and the place of humans in the world."[150] As Moore argues, this exploration leads to the understanding "of human limitations and the value of other, nonhuman life" as well as the temporal dimension provided by other human and non-human inhabitants of the space that have left their mark on it and create an awareness that "human ownership of the land is at best a short-term deal."[151] Re-

146 Ruland and Bradbury, *From Puritanism to Postmodernism – A History of American Literature*, 30–31.

147 Ruland and Bradbury, 31.

148 see Zilberstein, *A Temperate Empire*.

149 Ruland and Bradbury, *From Puritanism to Postmodernism – A History of American Literature*, 37.

150 Moore, *Ecological Literature and the Critique of Anthropocentrism*, 158.

151 Moore, 158.

maining in and engaging with one place allows Thoreau to deepen an experience of time that he sees as fundamentally different from his experience of living in town.

This contrast between different experiences of time, emerges most clearly in the juxtaposition of his own time frame, which has become attuned to the rhythms of his natural surroundings, and the intrusion of outside time frames – most prominently "railway time" via the Fitchburg railway line passing by the other end of Walden Pond: As Thoreau summarizes, his first summer in the woods was often spent in

> a revery, amidst the pines and hickories and sumachs, in undisturbed solitude and stillness, while the birds sang around or flitted noiseless through the house, until by the sun falling in at my west window, or the noise of some traveller's wagon on the distant highway, I was reminded of the lapse of time.[152]

This description already underlines a conception of time as personal and spent in contemplation, in harmony with the surrounding "nature" (the birds "flitt[ing] noiselessly through the house" seem almost cloyingly idyllic) until the intrusion of the outside world – in this case the "traveller's wagon" interrupts this "retirement" (to use Wordsworth's term here). Michael Jarret reads this sense of intrusion as capable of creating "alienation",[153] arguing, however, that Thoreau integrates the sounds of the railway into his soundscape, triggering reflection processes: "In Thoreau's writing on sound [...] we find moods becoming practices taking technological forms."[154] Thoreau's text constructs the sounds of the railway (rattling and whistling) as intrusive, but not in the sense of noise pollution, as the argumentative focus lies on the ways in which the sounds trigger thought processes and, thus, introduce a strange rhythm into a time-scape otherwise determined by the rhythms of "nature".

While the intruding sound of the outside world is, in this case, produced by a wagon, not the railway, the chapter lists the railway as the main reminder of a time frame belonging to the outside world. This impression is further supported by Thoreau's rejection of such strict conceptions of time as "days of the week" or "hours fretted by the ticking of a clock."[155] Instead, his descriptions suggest that he "amused" himself with housework[156] and developed a different concept of the pass-

---

152 Thoreau, *Walden*, 156–57.

153 Jarrett, "*Walden* + Railroad + Sound," 10.

154 Jarrett, 21.

155 Thoreau, *Walden*, 157.

156 Thoreau's relationship to housework has been widely discussed, with some critics of his project suggesting that he did not do his own laundry and was, therefore, a fraud. Some of this criticism seems to stem from the misunderstanding that Thoreau tried to cut himself off from all of society in an experiment of self-reliance and claimed to not have had any contact with society during the two years he describes in *Walden* – which is clearly based on a mis-

ing of time based on the growing of plants, the ripening of fruit and berries in the summer months. While this may sound like a romanticization of a life in tune with "nature", his comments on the sand-cherries which he tasted "out of compliment to Nature, though they were scarcely palatable"[157] indicate a much more mundane cataloging of many things growing in the area. As Richard B. Primack summarizes, Thoreau "made detailed observations of local flora and fauna [... and] had prepared tables precisely documenting the annual flowering times of three hundred plant species."[158] These tables, made "as background to a book he was planning but never wrote on the change of seasons"[159] were so detailed that Primack was able to use them as the basis of his own biological research on the ways in which climate change affected plant flowering times in the early anthropocene in *Walden Warming* (2014). While this part of Thoreau's work is by far not as well known as *Walden*, it shows that Thoreau's view of "nature" is, as Day argues, not necessarily comparable to the "high Romantic celebration of Nature", but "parallels what in England was a mid-Victorian engagement with a physical world increasingly defined by new geological and early evolutionary thought."[160] Thoreau sees "nature" not so much as benevolent or an entity to be mastered by the autonomous self, but rather as a systemic environment which humans must engage with out of a sense of self preservation. As Day argues,

> Thoreau loved nature, yet a primary implication of his writing about it is that while human beings must beware of over-exploiting it, they must do this in a sense less out of selflessness than out of self-love, out of a desire for survival of the species, lest the indifferent force of nature, drear, inhuman, unimaginably powerful in the long run, reacts simply by altering physical conditions on earth and, without concern, erases humanity and even life forms in general. [...] If destructive exploitation of natural resources is one manifestation of self-love on the part of humanity, of egotistical mishandling, then self-love may equally, though differently, express itself in the protection of the natural world.[161]

It is this understanding of living with and protecting one's natural environment from human exploitation as a form of personal growth that made Thoreau's un-

---

reading (or rather no reading) of the book. There is, however, also an underlying element of maliciousness in these comments that is incommensurable with feminist discussions on the distribution of care work. As Rebecca Solnit has argued, there are few writers who have had their laundry arrangements discussed in quite as much depth – often in an attempt to discredit Thoreau altogether. (Solnit, "The myth of 'Thoreau's mother did his laundry' will not die.")

157 Thoreau, *Walden*, 159.

158 Primack, *Walden Warming*, 5.

159 Primack, 5.

160 Day, *Romanticism*, 195.

161 Day, 195.

derstanding of "nature" particularly accessible to emerging environmentalist movements in the second half of the twentieth century, leading to a pop-cultural reframing of his work as ecological rather than sociological. Historically, Thoreau's engagement with inhuman nature is on the one hand impacted by a transcendentalist mindset focused on the observation and contemplation of his surroundings as a means of personal growth – a view that overlaps, in many ways, with the English Romantics' view of the relationship between poetic genius and "nature" as a source of poetic inspiration. On the other hand, his meticulous scientific observations also seem to run counter to his glorification of timeless "revery" quoted above. His interest in measuring his natural surroundings reaches back to the roots of Transcendentalism, not just in English Romanticism but also in Empiricism. It is not the concept of time or even the measurement of time that Thoreau, thus, seems to resist, but rather, as the rest of the "Sound" chapter suggests, the privileging of human time (measured by clocks and calendars) versus what might be described as non-human time, or perhaps "plant time", the rhythms of plant growth, seasons and the weather that Thoreau was attempting to capture in his notes, and to an extent also in *Walden*. As Russell West-Pavlov argues in *Temporalities* (2012),

> [t]he recent history of time since the Enlightenment has evinced a progressive narrowing of the spectrum of temporal modes. The gradual streamlining of temporality down to universal linear time as the self-evident calibration of human existence has repressed and elided other possible temporal structurings of individual and global existence.[162]

Thoreau's observations about the temporal impact of the railway capture a pivotal moment in this streamlining process. His objection to clocks and the suggestion of whole days spent in the same spot resonates with the observations about the railway, and its association with commerce, in the next part of the chapter, precisely because it runs counter to and undermines the rhythms of "natural" time:

> The whistle of the locomotive penetrates my woods summer and winter, sounding like the scream of a hawk sailing over some farmer's yard, informing me that many restless city merchants are arriving within the circle of the town, or adventurous country traders from the other side. [...] timber like long battering rams going twenty miles an hour against the city's walls, and chairs enough to seat all the weary and heavy laden that dwell within them. With such huge and lumbering civility the country hands a chair to the city. All the Indian huckleberry hills are stripped, all the cranberry meadows are raked into the city. Up comes the cot-

162 West-Pavlov, *Temporalities*, 6.

> ton, down goes the woven cloth; up comes the silk, down goes the woollen; up come the books, but down goes the wit that writes them.[163]

The passage – again ending on a witticism about the shrinking of intelligence among his contemporaries – is full of an imagery of speed and the destructiveness of bulky objects. The timber is compared to "long battering rams" attacking the city walls. The "hills are stripped" and "all" the produce is "raked into the city". This is not an argument against making chairs or picking berries, but against the sheer scale of these commercial enterprises made possible by the railway as the link between city and country, or, more precisely "city merchants" and "country traders" who are all hell-bent on commerce. It is in this sense that the last sentence must be read, because if writers involve themselves in this accelerated world, producing books at the same speed as chairs and thus seemingly without much thought, "down goes the wit that writes them." The whole passage functions as an argument against the acceleration of commerce made possible by the railway. As in Wordsworth's sonnet, the reader is left with the impression that there is a focus on the connection between the railway and trade as well as progress, but Thoreau takes a different, more mocking tone that sometimes seems to be admiring of the railway ("I too would fain be a track-repairer somewhere in the orbit of the earth"[164]) and sometimes sarcastically commenting on the ways the railway has changed the perceptions of the people:

> When I meet the engine with its train of cars moving off with planetary motion, – or rather like a comet, for the beholder knows not if with that velocity and with that direction it will ever revisit this system, since its orbit does not look like a returning curve, – with its steam cloud like a banner streaming behind in golden and silver wreaths, like many a downy cloud which I have seen, high in the heavens, unfolding its masses to the light, – as if this travelling demigod, this cloud compeller would ere long take the sunset sky for the livery of his train; when I hear the iron horse make the hills echo with his snort like thunder, shaking the earth with his feet, and breathing fire and smoke from his nostrils, (what kind of winged horse or fiery dragon they will put into the new Mythology I don't know,) it seems as if the earth had got a race now worthy to inhabit it. If all were as it seems, and men made the elements their servants for noble ends! If the cloud that hangs over the engine were perspiration of heroic deeds, or as beneficent as that which floats over the farmer's fields, then the elements and Nature herself would cheerfully accompany men on their errands and be their escort.[165]

---

163 Thoreau, *Walden*, 160–61.

164 Thoreau, 160.

165 Thoreau, 161–62.

The passage invokes a "new Mythology" by turning the train into a mythical animal bent towards a position among the stars. The use of nature imagery and the train's progressive ascent could almost be read as admiration, but the main focus of this passage is, once again, speed ("velocity"), as the railway is compared to a comet, shooting off into an uncertain orbit. There are elements of the sublime in this passage, evoked by the sheer scale of the comparison as well as the danger of being stomped by hooves or killed by fire. However, the ominous insertion of "[i]f all were as it seems" undermines this impression and the second part of the exclamation highlights humanity's ignoble ends in making "the elements their servants". The narrator's position – again, he does not seem to include himself in the general category of "men" here – and the fact that this sentence is highlighted with an exclamation mark, draws attention to its critical function. There is a clear sense of regret in the juxtaposition of human ingenuity and a personified "Nature herself" that also had to resist the intrusion of the railway in Wordsworth's sonnet. This is perhaps the closest that Thoreau comes to identifying the railway as a driving element of the anthropocene, whose beginning is marked by the arrival of the steam engine, as Morton argues. However, as Haraway, who has argued that "Capitalocene" might be a more accurate name for this epoch,[166] Thoreau's arguments are not aimed at implicating all of humanity, focusing instead on the destructiveness of accelerated commerce.

It is precisely in its function as a vehicle for trade that the train inaugurates a new form of temporality which, in turn, impacts how people interact with time. While "the passage of the morning cars" is "hardly more regular" than the "rising of the sun",[167]

> [t]he startings and arrivals of the cars are now the epochs in the village day. They go and come with such regularity and precision, and their whistle can be heard so far, that the farmers set their clocks by them, and thus one well conducted institution regulates the whole country. Have not men improved somewhat in punctuality since the railroad was invented? Do they not talk and think faster in the depot than they did in the stage-office?[168]

Taken out of context, this passage could be read as a eulogy on the railway as a means of regulating time and driving progress. Even the idea of a "well conducted institution regulat[ing] a whole country" could be read in this way. Set in the context of *Walden* and the discussion of living deliberately and observing the rhythms of "nature" as a form of self-improvement, the idea of a whole country working like clock-

166 see Haraway, "Anthropocene, Capitalocene, Chthulhucene." See also my discussion of Haraway's criticism of the term "Anthropocene" in the previous chapter.

167 Thoreau, *Walden*, 162.

168 Thoreau, 163.

work emerges as a nightmare vision of control. It is in this light, that the questions about "improv[ing] in punctuality" and "talk[ing] and think[ing] faster" are designed to elicit a sense of doubt in the reader, whether these are actually developments for the better. As Mee argues, drawing on Reinhard Koselleck, the pace of "industrial change" based on a "succession of technological innovation" was so brisk "that it appears as a change in the temporal regime of human society and its relations with nature. One result was the 'tension' described by Koselleck, already apparent by the late eighteenth century, 'between society and its transformation on the one hand, and its linguistic processing and assimilation, on the other.'"[169] This change in the understanding of temporality predates the process of standardizing time that was triggered by the expansion of railway systems. As West-Pavlov argues, in the nineteenth century, "discrepancies between regional time increasingly posed a problem for rapidly growing global capitalism [...] At the local American level, for instance, major railway accidents with numerous fatal casualties were caused from the 1850s on by the discrepant time schemes used by various railway companies."[170] While time zones and standardized time were not implemented in most European nations until the 1880s,[171] the passage from *Walden* – first published in 1854 and describing Thoreau's life in the woods in 1845–46 – still resonates with a sense of this shift in the understanding of time, if only at a local level. "To do things 'railroad fashion'"[172] means to do them in an accelerated manner that, in spite of the somewhat ironic tone of the passage, also conveys a sense of unstoppable progress, as the warning "to get off [the] tracks"[173] does not seem entirely metaphorical in this context.

Beyond the long-term project of "synchronizing the various local times",[174] Thoreau describes the railway as an innovation responsible for a far-reaching change in how people interact with it (in an almost worshipping manner, regarding it as a "demigod" as Thoreau suggests) and with each other (talking and thinking faster than before its invention and, thus, presumably reserving even less time for contemplative self-improvement). The ironic tone of these descriptions veils Thoreau's inherent criticism of this process of acceleration which countermands his earlier observations about his own more organic engagement with time. In contrast to the city people and citified country people who set their clocks by the train and improve their punctuality, Thoreau's own understanding of what might be considered time well spent, after all, includes the transcendentalist concept of

169 Mee, "'All That the Most Romantic Imagination Could Have Previously Conceived': Writing an Industrial Revolution, 1795 to 1835," 230.

170 West-Pavlov, *Temporalities*, 20–21.

171 see West-Pavlov, 21.

172 Thoreau, *Walden*, 163.

173 Thoreau, 163.

174 West-Pavlov, *Temporalities*, 21.

"reverie". This sense of temporality presents a complete contrast to the emerging concept of "universal time [which] tends to suppress local differences as it draws the entire globe into a single unitary temporal system."[175]

West-Pavlov highlights the role of logistics, that is, "the accurate planning of the transport of goods, and the tracking of their movement"[176] as one of the key factors in the development of universal time, an observation which resonates with Thoreau's description of the movement of goods (timber, chairs, berries) to and from the city. To the current reader, a phrase such as "[a]ll the Indian huckleberry hills are stripped, all the cranberry meadows are raked into the city"[177] may resonate as a more ecocritical argument than Thoreau might have intended, but there is a clear sense of environmental impact in these passages, suggested by the shear scale ("[a]ll", "stripped" and "raked") and "velocity"[178] of what is then framed as a progress narrative – albeit, ostensibly, not for Thoreau himself, but by the people he describes once again in somewhat ironic tones: "So is your pastoral life whirled past and away. But the bell rings, and I must get off the track and let the cars go by."[179]

While Thoreau does not focus his argument on the impact of the railway on the landscape, the descriptions of commerce made possible by the construction of the railway resonate with the potential for over-exploitation and destruction. As Gan, Swanson, Tsing and Bubandt argue in *Arts of Living on a Damaged Planet*, "[a]nthropogenic landscapes are also haunted by imagined futures. We are willing to turn things into rubble, destroy atmospheres, sell out companion species in exchange for dreamworlds of progress."[180] The authors draw attention to the "unplanned effects" of industrial projects: "Instead of building toward a single future, many kinds of time swirl through the worlds shaped by modern *anthropos*."[181] Thoreau's text, perhaps more than Wordsworth's seems to be haunted by the ghosts of the emerging anthropocene as an entirely unpredictable future underlying the capitalist ventures he describes as well as the settler colonialist slave-holder project in the US and many other parts of the world, which "require[d] ruthless ambition – and the willingness to participate in great projects of destruction while ignoring extinction as collateral damage."[182]

The accelerated progress of the railway calls forth such a haunting of an imagined future in Thoreau's description, as it does not only envision the actualities of the railway as a vehicle of trade, but also its impact on human society and the landscape

175 West-Pavlov, 22–23.
176 West-Pavlov, 23.
177 Thoreau, *Walden*, 161.
178 Thoreau, 161.
179 Thoreau, 167–68.
180 Gan et al., "Haunted Landscapes of the Anthropocene," G2.
181 Gan et al., G8.
182 Gan et al., G7.

alike. In this view "[m]inor forms of space and time merge with great ones"[183] as the impact of human beings on the planet in the anthropocene is cumulative, as well as accelerated "outpac[ing] all other rhythms of life."[184] Thoreau's juxtaposition of his own time frame of contemplative reverie and attempting to live in harmony with "nature" with the accelerationist project of building railways for commerce conjures the ghosts of future destruction and, thus, aligns itself with anthropocene readings. These are, as I have argued above, not so much based on the direct impact of steam power, as the long-term effects of releasing large amounts of CO^2 into the atmosphere could not yet be recognized, but rather on the ramifications of steam-powered travel and transport as ways of changing humanity's relationship to the non-human natural world. From a current point of view, Thoreau's *Walden* can, thus, be read as an early contribution to an ecocritical school of thought based in specific situated knowledges, but one that also needs to be contextualized as a political engagement with its historical context and read as distinct from its afterlife as a key text of American ecocriticism and a pop-cultural icon.

As I have argued above, to read Wordsworth's, Baillie's and Thoreau's text in the context of contemporary anthropocene criticism must also entail a focus on perspective, and the specific positionality from which these texts frame their relationship with non-human nature. I will address the question of perspectives further in the concluding part of this chapter, as it emerges in the context of approaches to gendered positionality and accessibility in the wake of innovations in steam-powered travel.

In Conclusion: Gender, Access and Steam-Powered Travel

In my discussion so far, I have taken William Wordsworth's and Henry David Thoreau's arguments concerning the introduction of the railway and how it shaped their understanding of space, time and the engagement with "nature" at face value as they were presented within the discourse of their time and, thus, privileging male subject positions as universally representative (Wordsworth) or situated in their own male-encoded embodied experience (Thoreau). In describing the impact of and engagement with the railway, both writers are also mainly focused on spaces associated with male-dominated areas of occupation such as farming, factories, construction work and trade. Moreover, their experience of travel is shaped by their own gendered perspective. Joanna Baillie's poem presents an outlier in this grouping, as it shows women's engagement in leisure travel as part of a mixed group of travelers. While there were, of course, women travelers and hikers in the Romantic

183 Gan et al., G5.

184 Gan et al., G5.

period, their experience of getting about was often different from men's, based on cultural conceptions of propriety, familial obligations or monetary limitations. Commerce, condemned as the root of all evil and the driving force behind the railway in Thoreau's and to an extent also Wordsworth's text, was not only firmly in the male domain, it also led to a further association of steam power with male domains from coal mines, via the construction of railroads and factories to trade on a scale that was only made possible by steam-powered travel.[185] Women rarely feature in these historical narratives, and class and disability emerge as intersecting categories of marginalization in this context, further limiting (or in rare cases also opening up) access to specific spaces, but also, for instance in the case of working-class women, adding time constraints that severely curtailed their access to aesthetic experiences.

Where gender emerges in the texts discussed above, it often does so in the form of association, or the inscription of gendered features. In Wordsworth's sonnet, for instance, a feminized "nature" is presented in opposition to the non-human, mechanical intrusion of the railway and while this entity is imbued with agency, it is neither human, nor ultimately likely to win the last stance against this "rash assault" that the speaker is demanding it take.[186] Thoreau's *Walden* paints a more complex picture of a society that also includes women, but its production and dissemination of knowledge in writing is rooted in the situated perspective of Thoreau's own experience and, therefore, tends to foreground activities traditionally associated with male-connoted strength and freedom of movement. While Thoreau's stance resonates with the feminist conceptualization of situated knowledge,[187] gendered inequalities are not among the social and political concerns he addresses in his discussion. Baillie, the only woman writer I have discussed in this chapter, occupies a middle ground, pursuing similar literary strategies to Thoreau in presenting a specific kind of situated knowledge, but also, as Wordsworth, imagining her speaker as an observer outside of the text, rather than a female traveler claiming space in her own right. This curious gesture describing the spatial specificity of a journey while the speaker positions herself as an observer who sees but does not take part in it may be rooted in the complex situation of female travelers and travel writers in the early nineteenth century who were, as Anne Rouhette summarizes,

185 Gidal reads steam-powered transportation not so much as the cause, but rather as a "necessary condition of the industrial revolution – at least in the British experience" (Gidal, "Industrial Transport and Political Economy in Blackwood's Edinburgh Magazine," 280.): "the network of roads, canals and coastal navigation expanded to meet the needs of a growing economy and laid a necessary path for the large-scale adoption of coal, whose transport from mines to urban and manufacturing centers motivated the construction of railways long before the network consolidation of the 1830s and 40s." (Gidal, 281.)

186 Wordsworth, *Kendal and Windermere Railway – Two Letters Reprinted from the Morning Post*, 4.

187 see Haraway, "Situated Knowledges: The Science Question in Feminism and the Privilege of Partial Perspective."

"affected by a suspicion of sexual impropriety which also concerned their very status as writers, since authorship was associated with authority, considered as exclusively masculine."[188] Distanced perspectives, thus, often emerge as concrete strategies in women's writing about travel. In Baillie's poem, the speaker's distance from the journey she describes allows her to take up an ungendered poetic voice imbued with authority which would not have been available to female travelers at the time, limiting the overt appearances of the speaker in the text to the expression of personal aesthetic opinion and gratefulness. This construction allows the speaker to present a view of the democratization of travel that is at once personal and rooted in the juxtaposition of different arguments.

For many nineteenth-century women writers experiences of steam-powered travel oscillated between a sense of freedom conveyed by an entirely novel experience of velocity and a mechanical power that seemed to liberate them from the restrictions of a world otherwise largely controlled by men and the strict social rules regulating women's lives which barred them from working and traveling freely in the same manner as men did.[189] As Anna Despotopoulou argues in *Women and the Railway, 1850–1915* (2022), the separation of spheres increased over the course of the nineteenth century, so that by the 1850s "Victorian gender ideology [...] imposed restrictions through discourses which constructed the spaces of the railway as topographies of patriarchy, fear, or manipulation."[190] These associations were, perhaps, not quite as well-established for the Romantics, but gendered stereotypes of women's constitutional frailty and fearfulness still resurface in some women writers' engagements with steam-powered travel in the first half of the nineteenth century. Mary Abdy's poem "The Railway Tunnel", composed between 1830 and 1850, already uses the common feelings of unease of travelling underground "[b]orne by the wondrous power of steam"[191] to construct an allegorical argument about faith and perseverance:

When dreary gloom our road invades,
He bids our hearts to thrill in lightness,
And guides us through the deepest shades,
To greet the day's unclouded brightness.[192]

This positioning of the titular railway tunnel, whose presence in the poem is tenuous to begin with, as entirely allegorical to move the poem towards a discussion of life's

188 Rouhette, "Gender Stereotypes in Some Romantic Travelogues (and How to Use Them)," 7.

189 see Despotopoulou, *Women and the Railway, 1850–1915*.

190 Despotopoulou, 2.

191 Abdy, "The Railway Tunnel," 185.

192 Abdy, 186.

ups and downs and the comforting faith in a "heavenly guide" is only possible because of the commonness and consequent relatability of the experience of traveling "[w]ith startling speed through regions dreary", which leave "[s]ome panic-struck and fearful" and some "languid, spiritless, and weary."[193] Born in 1797, Abdy had her poetry privately printed from the 1830s onwards, suggesting that by that time traveling through a railway tunnel was a common enough experience to be used in this kind of allegorical construction.

As I have argued above, the inevitable engagement with the railway changed people's perceptions of time and space and, to an extent, what it meant to be human in a confrontation with new technologies on the one hand and the non-human natural world on the other. As Despotopoulou shows, these shifts were perhaps even more pronounced for women in the nineteenth century, who came into contact with a plethora of new experiences by way of the more democratic possibilities of railway travel. Despotopoulou's project on "women in transit" reads the railway as "an ambiguously gendered space that highlights the artificiality of the private/public divide, giving prominence to woman's impulse to traverse boundaries, not only physically but also mentally and emotionally."[194] The timeframe of her study lies somewhat later then the Romantic texts I am interested in here, while also focusing on the railway as a built environment rather than an intrusion into a "natural" space. As Despotopoulou argues, in the Victorian era, the train and the railway station emerged as newly conceived spaces of transit "in which the rules of circulation for men and women had to be formulated anew, balancing social conventions concerning gender behaviour with the emerging demands of travel, industry, and work, served primarily by the railway."[195] Rather than focus on the ways in which travelers engaged with these newly built spaces, my emphasis throughout this chapter has been on the ways in which steam-powered travel is conceptualized in Romantic writing – especially in its emergence in texts that focus on the writer's own position in and towards "nature". Gendered expectations around travel (or rather the possibility to travel in the first place) complicate these discussions as they impacted the ways in which men and women writers of the Romantic period moved through space as well as their poetic conceptualization of these movements. As Rouhette argues,

> [i]n spite of highly successful travel books [...] travel writing remained gendered masculine in the early 19th century and beyond. This is largely due to the fact that the exploration and conquest of space were associated with men and antithetical to the idea of the private sphere to which women were increasingly relegated as

193 Abdy, 185.

194 Despotopoulou, *Women and the Railway, 1850–1915*, 2.

195 Despotopoulou, 3.

> the 18th century wore on and as conduct-book writers like James Fordyce stressed the domestic circle as woman's natural environment.[196]

Aesthetic, stylistic or thematic differences in travelogues by male and female writers of the Romantic period described by Rouhette (who draws on previous work by Nicole Pellegrin and Sara Mills) are, however not as clearcut as one would expect based on the restrictions imposed on women's movements in public spaces and tend, moreover, to run the risk of essentialist reductions to physical differences between bodies read as male or female. Rouhette summarizes critical attempts to distinguish between the poetics of male and female writers, which set up a "system of binary oppositions [...] opposing detail and generalisation, observation and reflection, immediacy and distance, subjectivity and objectivity" to establish a male and a female way of writing based on "the writer's sex"[197], but concludes that, in the examples she discusses, it is

> impossible to differentiate between an essentially "masculine" or a "feminine" mode of travel writing, insofar as expectations of femininity and of feminine writing have been so internalized by Romantic women writers that it is virtually impossible to distinguish between a naturally, quintessentially feminine type of writing, whatever that may be, and a conscious or unconscious conformity to the prevailing conception of what true female "nature" should be.[198]

Where femininity is, thus, inscribed on bodies based on how they are perceived, it finds an expression in the form of gendered expectations and conventions and their performance, which are, in turn, shaped by a writer's specific positionality. Rouhette reads the Shelleys' *History of a Six Weeks' Tour* (1817) as an example of the ways in which the two contributors to this journal adopted ostensibly gendered voices in their descriptions of both their experiences and their surroundings, thus not only making it impossible to distinguish between the two voices but also drawing attention to the adoption of specifically gendered forms of discourse as a construction in the text: "[B]y playing on expectations of masculine and feminine discourses which the very genre of travel writing crystallises [...] the Shelleys' *History* [...] indirectly question[s] ideological assumptions about the representations of masculinity and femininity prevalent in British society at the time."[199] Rouhette reads this as part of the Shelleys' political program, for which they "manipulate gender stereotypes to express political views while deflecting potential criticism."[200]

---

196 Rouhette, "Gender Stereotypes in Some Romantic Travelogues (and How to Use Them)," 7.

197 Rouhette, 12.

198 Rouhette, 4.

199 Rouhette, 28.

200 Rouhette, 28.

Mary Shelley's novel *The Last Man* also employs this type of multilayered gender performance, complicating a narrative perspective that is ostensibly identified as male. While I will discuss the novel in more detail in the next chapter on "Apocalypse", I would like to use an example of this strategy from the novel here, to draw attention to this effect in the context of travel writing. In this passage, the autodiegetic narrator describes the funeral procession that carries the body of Lord Raymond (Lord Byron's alter ego in this *roman à clef*) from Constantinople to the site of his grave near Athens:

> Behold now the melancholy train cross the flats of Thrace and wind through the defiles, and over the mountains of Macedonia, coast the clear waves of the Peneus, cross the Larissean plain, pass the straits of Thermopylae, and ascending in succession Oerta and Parnassus, descend to the fertile plain of Athens.[201]

The "train" in this case is a procession of riders and carriages, not an actual train and the speed of the procession is determined not only by the occasion but also by the vehicles and comfort of the travelers. Distance and stately speed are marked by both active verbs and geographical name-dropping, evoking Greek history and mythology to underline the role of Raymond as an important figure in the Greek war of independence and the sad occasion of conveying his body to Athens for his burial. The strain the drawn-out time of this slow procession puts on the travelers is curiously gendered in the passage. As the male narrator observes,

> [w]omen bear with resignation these long drawn ills, but to a man's impatient spirit, the slow motion of our cavalcade, the melancholy repose we took at noon, [...] the monotonous recurrence of day and night, unvaried by hope or change, all the circumstances of our march were intolerable.[202]

The juxtaposition of men's and women's different experiences of travel is further embedded in a generalization of their emotional reactions of "resignation" and "impatience", thus, seemingly functioning as a re-inscription of these stereotypes on the part of the narrator, who seems to be expressing his own view on the "intolerable" situation. In voicing his discontent of the "long drawn ills" of traveling at "slow motion" in this manner, the narrator explicitly frames it within a gendered comparison that privileges women's experience of and emotional reaction to it by setting their experience as the standard, which women are used to and from which men deviate. The narrator's comparison generalizes the perspective of both "women", via

201 Shelley, *The Last Man*, 208.

202 Shelley, 208.

their common experience of having to travel at a slow pace, and men, whose impatient spirit can not tolerate these circumstances, thus turning the experience of this specific journey into a universal one. Verney does not adopt a female voice or report on a specific female perspective – even though there is one immediately available to him – that of his sister Perdita – Raymond's widow, who "shut up in herself, spoke little" and "refused communication or sympathy", thus modeling "resignation" in her actions.[203] Instead of drawing on Perdita's experience, as he does in other parts of the text, Verney's description of her withdrawal, if not inconceivable for a grieving widow, nevertheless also functions to explicitly distance her from his gender-focused thoughts about the cumbersome journey.

Verney's view is further complicated by the complex narrative construction of the novel, which embeds this comment on women's "resignation" to slow travel in an Englishman's autodiegetic narrative of the imagined last years of the twenty-first century which, as the novel's frame narrative explains, is a reconstruction of a prophecy discovered on a selection of leaves discovered in the "gloomy cavern of the Cumaean Sybil" by an unidentified autodiegetic frame narrator who "visited Naples in the year 1818."[204] As in *Frankenstein*, which weaves its thoughts on humanity, science, reproduction and authority through a complicated web of narratives, *The Last Man*'s semi-fictional frame narrative also complicates the different perspectives on "nature", gender, politics and poetics the novel voices, but the novel also flattens temporal distinctions by merging the (fictionalized) voices of a 6th-century BC seer, an early 19th-century writer and a 21st-century traveler to describe a world that is largely based on Mary Shelley's own time period. While there are, of course, genre-specific differences between travel writing and the novel, *The Last Man*'s multilayered narrative construction embeds Verney's comments on traveling within a similar framework of gendered narrative strategies as the travelogues described by Rouhette, which "indirectly question ideological assumptions about the representation of masculinity and femininity" and "manipulate gender stereotypes to express political views."[205] What emerges from these examples is, thus, not so much a confirmation of the existence of "male" and "female" strategies of travel writing – much less of noticing –, but rather an understanding of how Romantic writers (men and women, as Rouhette shows) were aware of and used these strategies to achieve specific effects. As I have shown for Baillie's example above, perspective also emerges as a complicated construction in women's writing on steam-powered travel, as the text attempts to negotiate the conflicting exigencies of access and aesthetics.

203 Shelley, 208.

204 Shelley, 6.

205 Rouhette, "Gender Stereotypes in Some Romantic Travelogues (and How to Use Them)," 28.

The critical discussion of women travelers in the nineteenth century is frequently focused on the ways in which the definition of different spheres in the common cultural discourse limited women's access to specific experiences. Here, I would like to come back to the ways in which this question of access shapes what could be perceived and what could be written about. Beyond the socio-cultural limitations of access to certain spaces and activities, the texts on steam-powered travel I have discussed in this chapter also draw attention to the material conditions of access to certain spaces. Baillie's poem most prominently shows that for certain groups of people the steamer would have been their only possibility of travel and, consequently access to an aesthetic experience of a landscape that would otherwise have been unreachable for them because of a complex set of monetary, temporal and socio-cultural restraints. The steamer, in this sense functions as a kind of aesthetic prosthesis that extends the range of function and accessible spaces and experiences for the travelers. Baillie's description of steam power as magic echoes other contemporary opinions on the agency of machines. As Bridget Marshal argues,

> [t]he way that machines moved without apparent human intervention or oversight was often seen as uncanny, suggesting that machines had minds of their own. When Dorothy Wordsworth came upon one of Watt's steam engines in the Scottish countryside in 1803, she expressed her feeling that 'it was impossible not to invest the machine with some faculty of intellect; it seemed to have made the first step from brute matter to life and purpose'.[206]

Understood in this sense, steam power did not only broaden the possibility of human experience, but its prosthetic agency also raised uncomfortable questions about the fundamental understanding of agency, intellect and, by extension the human and its Others that continue to resonate in the context of cybernetic organisms, lifelike robotics and the possibility of artificial intelligence today. These discussions often align themselves with other forms of marginalization in the feminist Posthumanities, as they probe at the limits of who or what is considered "human" in mainstream discourses. Dorothy Wordsworth's observations about the agency of the steam engine could, consequently, be reframed not as an expression of uncanny horror, but as a way of drawing attention to the ability of technological innovations to raise questions about human agency. Baillie's reading of the steam-vessel as a "laborer" seems to reflect a similar understanding of technological agency that aligns itself with the steam-vessel's role in making the landscape accessible.

To argue for walking as the only aesthetically and personally valid way of experiencing the countryside as both Wordsworth and Thoreau do (if for slightly different

206 Marshall, *Industrial Gothic*, 127.

reasons), thus, emerges as not only a fundamentally classist, but also an ableist position. Baillie's poem does not explicitly embrace accessibility (in the sense the term is used in the context of disability studies and activism today), but the assurance that the passengers of the steamer "[n]ow seated at ease may glide along"[207] Loch Lomond stands in direct opposition to Wordsworth's insistence of only being able to enjoy the full aesthetic experience of the landscape while traversing it on foot.

John Harden: "Dorothy Wordsworth in a Wheelchair" (1842)

Abbot Hall Art Gallery / Bridgeman Images

William Wordsworth's somewhat flippant comment on Dorothy Wordsworth "PS [...] Your Aunt Dorothy goes on as usual, she is still whee[led] about in her carriage twice a day"[208] (probably the wheelchair from John Harden's sketch – which also replicates ableist narratives of suffering and uneasy compassion), is indicative of the ways in which her access to "nature" would have relied on mobility aids and the help of others at a time (in the 1840s) when her brother was propounding the necessity of walking for a proper aesthetic experience of the landscape. The critical discussion of Dorothy Wordsworth's writing tends to contrast her work as a "nature writer" in her earlier years, while she was able to go on vigorous walks through the

207 Baillie, "Address to a Steam-Vessel," 41.

208 From a postscript to an undated letter by William Wordsworth to his nephew, probably written in December 1845 (Wordsworth and Wordsworth, *The Letters of William and Dorothy Wordsworth – The Later Years, Part II*, 728.).

countryside, with ableist dismissals of her later work, due to her limited access to what was perceived as "nature". Moreover, as Polly Atkin and Jo Taylor conclude in a conversation about Atkin's book *Recovering Dorothy* (2022), an ecofeminist discussion of Romantic writing needs to take into account how an understanding of "writing nature" is often limited to very specific perspectives, raising ecofeminist questions about "who gets to write nature and how" and also "who is included in ecological thought."[209] As I have argued for the botanical work of Elizabeth Kent in the chapter on "Anthropocene", a narrow understanding of "nature" would have excluded domestic experiences and engagements with non-human nature in and around the house, automatically precluding women writers who wrote about potted plants (as Kent did) or birds and plants and memories of earlier experiences she was still able to engage with from her sick bed (as Dorothy Wordsworth did). Atkin draws attention to this distinction between valid and marginalized perspectives in the critical engagement with Dorothy Wordsworth's work before her illness:

> Dorothy Wordsworth is praised as a nature writer, and as a woman who walked and climbed. She is admired for her physical vitality as well as for her writing: for walking long distances, for stamina, strength, and a certain wildness. Her worth as a writer has become tied to her physical capacity by book after book which repeats these connections.[210]

This critical discussion does, as Atkin points out, not include the intersecting factor of her disability, nor does it address the ways in which she thought of and managed to engage with the non-human natural world while largely unable to leave the house or even her bed from the 1830s onward.[211] While, as Atkin points out, much of her day-to-day writing from this period was destroyed by well-meaning relatives bent on preserving a specific legacy that did not include sickness, Dorothy Wordsworth's poem "Thoughts on my sick-bed" (1832) gives a glimpse of her "hidden life" that also engages with William Wordsworth's poetics:

> No prisoner in this lonely room,
> I saw the green Banks of the Wye,
> Recalling thy prophetic words,
> Bard, Brother, Friend from infancy!
> No need of motion, or of strength,
> Or even the breathing air;

209 *Recovering Dorothy.*

210 Atkin, *Recovering Dorothy*, 11.

211 see Atkin, *Recovering Dorothy*.

– I thought of Nature's loveliest scenes;
And with Memory I was there.[212]

The poetic engagement with "Nature's loveliest scenes" is filtered through a double process of reflection here, as the poem first recalls her brother's "prophetic words" that "Memory" is a way of "being there" and then thinks of the remembered scenes to put this process into action. The poem, however, also complicates William Wordsworth's poetics of "emotion recollected in tranquility", imagining a writer whose experience of "Nature's loveliest scenes" comes from "Memory", because the distant natural world itself is no longer accessible to her. While the text itself repeats a strict separation of inside and outside, differentiating "this lonely room" and "Nature's loveliest scenes", it also argues that "Memory" can traverse this boundary and allow the writer to engage with the natural world in spite of lacking "motion" and "strength". The poem meta-poetically comments on the source of this notion – William Wordsworth's own poetics of "emotion recollected in tranquility",[213] but there is an important difference in the critical perception of Dorothy Wordsworth's "couch of rest" – which is seen as a symbol of illness and confinement – and the "couch" William Wordsworth imagines lying on "in vacant or in pensive mood" to see the daffodils dance before his "inward eye" in "I wandered lonely as a cloud".[214] While William Wordsworth's poem is understood as an expression of poetic reflection of an earlier experience of the natural world, Dorothy Wordsworth's writing is reduced to the reflection of her illness in almost every reading of this poem. This difference clearly does not stem from the piece of furniture itself – it might, for all intents and purposes, have been the same couch that brother and sister lay on to contemplate their memories of flowers and "Nature's loveliest scenes" at different moments in their lives – nor does it reside in the action of lying down and remembering an experience of the natural world, but rather in whose experience is valued as poetically relevant in the first place.

Gender and disability work as intersecting modes of exclusion in this context, curtailing private, public and critical perspectives on Dorothy Wordsworth's later years. As Atkin argues in *Recovering Dorothy*, "Dorothy after 1829 lives in the alternative time of crip time, travelling not only through time but also through place in her poetry, and in her vivid, visionary memory."[215] Crip time exists on both a systemic and a personal level, but individual perceptions of crip time and crip engagements with spatiality often run counter to official, standardized notions of tempo-

212 Wordsworth, "Thoughts on My Sick-Bed."

213 Wordsworth, "Preface to Lyrical Ballads," 361.

214 Wordsworth, "Daffodils," 383.

215 Atkin, *Recovering Dorothy*, 123.

rality and spatiality.[216] While steam-powered travel disrupts notions of temporality, the discussion of these disruptions emerges as individual resistance to acceleration and standardization processes in Wordsworth's and Thoreau's arguments – partly because attempts at universalization of individual experiences fail to acknowledge the problem of accessibility and partly because their situated knowledge does not include this specific positionality. Baillie's argument for the democratization of access to the landscape through steam-powered travel, by comparison, acknowledges the desirability of access, but is also aware of incongruencies and challenges arising from conflicting approaches to technology and aesthetics. While none of the three texts that I have discussed here engage with the problems of pollution and dwindling resources associated with the use of fossil fuels in more recent anthropocene discourses, these issues further complicate problems of access, as they draw attention to the question of who is included and excluded from discussions around climate crisis on the basis of various forms of marginalization.

Accessibility, in this sense, does not just encompass the question of who is granted access to specific spaces and experiences, but also, by extension, whose experiences are considered valid and meaningful, as I have argued with regard to Dorothy Wordsworth above. These questions have ramifications that continue to echo through ecocritical engagements with Romantic writing as well as an understanding of the ecological movements that arose from early nineteenth-century conservationism and continue to influence official engagements with what is perceived as "nature" and by whom and in what ways different cultures interact with the non-human natural world today. While I will further address these questions in my Conclusion, the next chapter – "Apocalypse" – presents a further exploration of the ways in which gender emerges as a complicating factor in poetic and narrative engagements with the relationship between "man" and "nature" in the context of climate catastrophe. The texts I analyze – Byron's "Darkness", Mary Ann Browne's "A World Without Water" and Mary Shelley's *The Last Man* – negotiate the possibility of positionality in apocalyptic scenarios that resist attempts at being aesthetically framed as sublime. In this context, distinct strategies of positionality emerge as male and female writers struggle to position themselves within an increasingly secularized universe.

216 see McRuer, *Crip Times*.

Chapter 3: "Apocalypse"

Introduction: Climate Apocalypse and the Sublime

> I awoke in a painful agony – for I fancied that ocean, breaking its bounds, carried away the fixed continent and deep rooted mountains, together with the streams I loved, the woods, and the flocks – it raged around, with that continued and dreadful roar which had accompanied the last wreck of surviving humanity.[1]

Taken out of the context of Mary Shelley's 1826 novel *The Last Man*, this passage might be read as prophetic of the emerging twenty-first century realities of climate catastrophe with its very real threats of melting polar caps, rising sea levels, flooded coastal regions and disappearing islands states.[2] The nightmarish images of an end of the natural world conjured by Shelley's lone surviving protagonist have become imbued with meaning to an extent that Rebecca Richardson describes as uncanny,[3] moving far beyond anything that the novel imagined for the late 2090s, but also holding a mirror up to a period that is now, as I have argued above, understood as crucial to the emergence of the anthropocene. The passage draws attention to the manner in which recent representations of climate catastrophe in the anthropocene resonate with echoes of Romantic discourses around "nature", its meaning to humanity, and the possibilities of its destruction. Its dreamlike construction, however, also draws attention to the problem of imagining a subject position from which such a devastating scenario might be described.

Set in the far future, the text points backward to its own time, as well as forward towards an anthropocene perspective which is steeped in Romantic echoes, as I have

1 Shelley, *The Last Man*, 447.

2 see IEP, *Ecological Threat Register*.

3 Richardson, "The Environmental Uncanny: Imagining the Anthropocene in Mary Shelley's *The Last Man*," 1063.

argued in the "Anthropocene" chapter above. This evokes Romanthropocene connections for readers and critics alike. As Richardson argues, to read early anthropocene texts from the current moment also raises the "question of how we can or should read our own concerns in texts that could not have anticipated them" without running the risk of reducing interpretations to an enterprise in "combing the past for the catalysts behind the fossil-fuel-powered global system we find ourselves in."[4] There is, however, a fine balance between Richardson's warning against reading present concerns *into* past texts and John Beer's more general assertion that readings of apocalyptic texts somewhat automatically acquire additional meaning, because they tend to be framed within the interpretive scope of threats relevant to the critic's current moment.[5]

Instead of claiming that the apocalyptic texts I read in this chapter are *about* climate crisis, I would like to suggest that anthropocene concerns emerge in unprecedented and unexpected ways in early anthropocene texts, warranting an historical extension of Mark Bould's argument in *Anthropocene Unconscious* that "the Anthropocene *is* the unconscious of 'the art and literature of our time'."[6] In order to visualize these underlying concerns, Bould suggests a climate change-related queering of our reading practices, asking "what happens when we stop assuming that the text is not about the anthropogenic biosphere engulfing us? What if all the stories we tell are stories about the Anthropocene?"[7] While Bould's interpretive range is directed at the ways in which anthropocene concerns pervade the literary and cultural production of the current moment, I would like to consider early anthropocene texts as poetic and narrative engagements that resonate with emerging anthropocene concerns in ways that are, perhaps, not always straightforward and easily discernible, grappling with problems of positionality and representation that meta-textually draw attention to climate catastrophe as an unconscious threat.

In this manner, I read the three Romantic apocalyptic texts I discuss in this chapter – Byron's "Darkness" (1816), Mary Ann Browne's "A World Without Water" (1832)

4 Richardson, 1064.

5 Beer, "Romantic Apocalypses," 109. Beer's article, published in early 2001 and addressing the Romantics' engagement with apocalypse from an angle unrelated to climate crisis, discusses connections to recent events (from his perspective), including fears of the millennium bug, as well as millennial fears emerging around earlier turns of the century.

6 Bould, *The Anthropocene Unconscious – Climate Catastrophe and Culture*, 15. Bould's emphasis is directed at Amitav Ghosh's argument in *The Great Derangement* that climate catastrophe is absent from literary fiction (if not genre fiction) of our time (see Ghosh, *The Great Derangement*.), but it also draws on Fredrik Jameson's argument in *The Political Unconscious* (1981), which disputes the idea that literary creation might take place and be considered outside of its political context (see Jameson, *The Political Unconscious*.). See also my discussion of objectivity and situated knowledges in the "Anthropocene" chapter above.

7 Bould, *The Anthropocene Unconscious – Climate Catastrophe and Culture*, 17.

and Mary Shelley's *The Last Man* (1826) – as literary engagements with pressing questions pervading early anthropocene discourse, for instance the ways in which drastic technological and cultural changes had begun to instigate major shifts in the ways in which the Romantics understood and engaged with the relationship between "man" and "nature" and how these questions could be framed aesthetically and from whose subject position. More than simply asserting that these texts are *about* climate crisis, I would like to suggest that they reflect the major positional and aesthetic shifts necessary within Romantic poetics to critically frame the idea of climate catastrophe, untethered from divine providence or the possibility of aesthetic sublimation. This critical engagement with the underlying aesthetic concerns is a necessary approach to Romantic writing engaging with the aftermath of the 1815 eruption of Mount Tambora, precisely because the Romantics themselves were unaware of the original event and their poetic and fictional engagements with its aftermath are frequently vague and speculative as to the causes of and connections between the particular catastrophes they engage with. In a sense, Tambora presents a parallel to climate change, as for both the causes remain invisible – or unconscious – but both also produce highly visible effects that are frequently connected in seemingly inexplicable ways.

The Tambora crisis emerges in different ways in early nineteenth-century literature, influencing socio-political as well as literary and aesthetic discourses beyond the immediate knowledge of the Romantics and, thus, largely unrecognized or seen as unconnected by them. In anglophone popular culture, Tambora is now frequently referenced as having caused "the year without a summer" (1816), but consequences of Tambora's eruption were more far-reaching, causing three years of unseasonably cold summers, crop failures and outbreaks of diseases such as typhus and cholera, which caused a staggering number of worldwide deaths in addition to the more than 100.000 immediate victims of the event in Indonesia.[8] As Gillen D'Arcy Wood observes, Tambora's eruption is the most recent volcanic event on a major scale and, therefore, offers a unique case study of the global impact of climate change:

> [F]acing multiplying extreme weather crises of our own, its eruption looms as the richest case study we have for understanding how abrupt shifts in climate affect human societies on global scales and decadal time frames. The Tambora climate emergency of 1815–18 offers us a rare, clear window onto a world convulsed by weather extremes, with human communities everywhere struggling to adapt to sudden, radical shifts in temperatures and rainfall, and a flow-on tsunami of famine, disease, dislocation, and unrest.[9]

8 see Behringer, "Die Tamborakrise."

9 Wood, *Tambora*, 8–9.

While the effects of Tambora's eruption were felt around the globe, the Romantics lacked the necessary modern scientific instruments and data to register a volcanic eruption thousands of miles away and make the connection to local weather phenomena.[10] Unrelated events, such as the unusual but coincidental increase in sunspots,[11] also led to fallacious connections, so that the uncoupling of the effects from the original cause of climate catastrophe resulted in a plethora of speculative engagements with climate crisis in Romantic literature after 1815.

The immediate connections not available to the Romantics are highly visible in retrospect and the Tambora crisis has, consequently attracted recent critical attention in the context of early anthropocene criticism interested in the Romantics' understanding of climate change. Higgins has analyzed in depth how the Byron-Shelley circle engaged with the inclement weather of the "wet, ungenial summer"[12] of 1816 spent in Switzerland and how their view of the Mont Blanc glacier and reading of contemporary theories around the newly conceived concept of "deep time" triggered reflections of climate change and apocalypse in their writing.[13] Both Higgins' *British Romanticism, Climate Change and the Anthropocene* (2017) and Reno's *Early Anthropocene Literature in Britain* (2020) offer in-depth discussions of the ways in which the Shelleys and Byron engaged with contemporary climate theories leading up to and during their stay at the Villa Diodati in the summer of 1816, and their accounts are highly useful for understanding the theoretical and material background to key texts such as Percy Shelley's "Mont Blanc" (1816), Byron's "Darkness" (1816) and Mary Shelley's *Frankenstein* (1818). My discussion here moves beyond the scope of these critical texts in several ways: most prominently in that I present a gender-focused reading of the ways in which early anthropocene climate texts envision a subject position in the context of immediate and extensive climate catastrophe. My readings are, consequently, more concerned with textual aesthetics than with the theoretical and material dimensions of the Romantics' immediate reactions to the Tambora crisis. To this end, I have also broadened the temporal scope of my readings beyond the direct aftermath of Tambora to trace Romantic engagements with ideas of secular apocalypse in the context of climate catastrophe after 1816 and, therefore, jux-

10 Wood, 7.

11 Jeffrey Vail draws attention to the prophecy of an "Italian astronomer [who] had predicted that the sun would burn itself out on 18 July" 1816 – a prediction that seemed to be substantiated by the high number of sunspots and cold temperatures of this summer (Vail, "The Bright Sun Was Extinguish'd," 183.).

12 Shelley, *Frankenstein*, 170.

13 Higgins, *British Romanticism, Climate Change and the Anthropocene – Writing Tambora*, 109 ff.

tapose Byron's "Darkness" with Mary Shelley's novel *The Last Man*[14] as well as Mary Ann Browne's poem "A World Without Water".

I argue that the three texts are not only thematically connected in that they imagine large-scale catastrophic events eradicating life on earth, but that they also poetically and structurally reflect the impossibility of framing such an event within the scope of the eighteenth-century aesthetic apparatus of the Burkean sublime. This inability to aesthetically encompass climate catastrophe is not only closely related to the problem of constructing an observing/poetic subject position in the face of an immediate extinction event. The breakdown of aesthetic structures also highlights problems of positionality in Romantic aesthetics that, as I argue throughout, become particularly visible in the context of early anthropocene climate change and climate catastrophe. Connections to the current climate crisis emerge not so much via the continuing threat of climate crisis but through common questions concerning the ways in which a subject position might be imagined in the face of multiple, related crises and, by extension, the resonances and residues of Romantic constructions of positionality in current engagements with climate crisis in the anthropocene.

While the three texts are connected by their interest in the ways in which climate catastrophe forces an engagement with the human subject's position and connection to the non-human natural world in an increasingly uncertain universe, gender emerges as a complicating category in this context. As I argue throughout, a Romantic subject position rooted in the concept of the aesthetic construction of the autonomous self was never available to women writers to begin with, making it necessary to root experience in the body, rather than an understanding of sublimation available through an experience of fusion with "nature". These concerns emerge in the three texts in different ways connected to their understanding of gendered positionality, their textual aesthetics, and questions of genre, but the texts also grapple in similar or comparable ways with the underlying problem of having to move beyond the ultimately comforting aesthetic experience of the sublime in the experience of an apocalyptic "nature" which denies the experiencing subject a confirmation of the autonomous self.

Burke's understanding of the sublime, on which the Romantics draw, follows a clear binary structure that distinguishes between beauty – associated with weakness and femininity – on one side and the sublime – related to masculinity, agency and strength – on the other.[15] Feminist resistance to gendered theorizations of

14 Heidi Scott also argues that *The Last Man* might be a better example than *Frankenstein* of a "post-Tambora text" with implications for the current understanding of climate change (see Scott, *Chaos and Cosmos*; see Murphy, "Apocalypse Not Quite.").

15 see Burke, *A Philosophical Enquiry into the Origin of Our Ideas of the Sublime and the Beautiful*; see Mellor, *Romanticism and Gender*, 85ff.

the sublime reaches back to the eighteenth century, as Mary Wollstonecraft contests Burke's view in *A Vindication of the Rights of Man* (1790), suggesting that the sublime originates from moral truth, whereas beauty is based on simplicity. Wollstonecraft's counterargument suggests that women are socially discouraged from pursuing "truth, fortitude and humanity" in their own right, and, as a consequence, pushed towards "neglecting morals to secure beauty" to become aesthetic objects instead of perceiving and acting subjects.[16] While attempts at reclaiming the sublime push against its dualistic construction of gender, the sublime in Burke's sense functions, as Bonnie Mann has argued, as a mode of Euro-masculine self-constitution,[17] with far reaching consequences for gendered aesthetics and the post-Romantic afterlife of the sublime as an aesthetic category. Both the Kantian and the Burkean sublime operate on two connected levels: that of the perceiving subject and that of the perceived object. For both the sublime and the beautiful, perception is pre-rational, but the sublime is associated with overcoming the pre-rational sentiment through the agency of the perceiving subject, whereas the beautiful does not possess agency in itself, nor does it produce a sense of agency in the observer. The necessity of agency in the sublime arises from the "sense of boundlessness and indeterminacy that threatens the definition of the subject"[18] as it is confronted with the unknown and also largely unknowable. This pre-rational experience of the sublime requires an experiencing aesthetic subject capable of being sublimated, that is, altered by the aesthetic experience and a sublime object capable of providing this experience. In Romantic aesthetics, this construction emerges as the experience of the autonomous self in its aesthetic fusion with and consequent mastery of "nature".

Burke's argument hinges on a distance created between the perceiving subject and the perceived sublime object that is either created by the fact that the object merely excites an "*idea* of pain, and danger", is "*conversant* about terrible objects" or "operates in a manner *analogous* to terror."[19] Burke argues that "When danger or pain press too nearly, they are incapable of giving any delight, and are simply terrible; but at certain distances, and with certain modifications, they may be, and they are delightful."[20] This distance is only available to the observing subject who is able to take up a position outside of the observed sublime object. One of the most frequently cited examples in this context is a volcanic eruption, as only the necessary distance

16 Wollstonecraft, "On Poetry," 47.

17 see Mann, *Women's Liberation and the Sublime: Feminism, Postmodernism, Environment*.

18 Mishra, *The Gothic Sublime*, 19.

19 Burke, *A Philosophical Enquiry into the Origin of Our Ideas of the Sublime and the Beautiful*, 36–37 (my italics).

20 Burke, 36–37.

of the viewer from the object – the volcano – allows for an aesthetic experience of the sublime.

The primacy of this understanding of the sublime in eighteenth-century aesthetics and beyond has far-reaching consequences for climate discourses as it provides a theoretical framework that hinges on the construction of a subject position outside of the immediate range of catastrophe. Morton's conceptualization of climate crisis as a "hyperobject" criticizes this understanding of distance between the observing subject and the observed object, arguing that there is no meta-language – no position outside of ecological and subjective situatedness in the anthropocene, from which humanity's impact on the planet could be discussed.[21] Morton's understanding of the sublime draws on Kantian phenomenology and a Romantic understanding of humanity's position as distinct from "nature":

> The classic image of Nature is the Romantic or picturesque painting of a landscape. There it is, over yonder – on the wall in the gallery. And it has over-yonderness encoded throughout it: look at those distant hills, that branch suggesting that we follow the perspective lines toward the vanishing point, and so on.[22]

While the sublime and "nature" can exist outside of and beyond the human subject position in eighteenth-century aesthetics and Romantic poetics, hyperobjects in Morton's sense, can not be aestheticized within the framework of the sublime because there is no possibility of an external subject position from which to consider the anthropocene.

My main argument in the following is based on this Romanthropocene connection between the complexities of constructing a subject position in sublime aesthetics and Morton's understanding of the immediacy of climate crisis as a hyperobject, arguing that early anthropocene texts reacting to an emerging sense of climate catastrophe already grapple with the problem of situating an observing subject outside of the observed apocalyptic events and, as Morton argues, "situatedness is now a very uncanny place to be."[23] This problem emerges in different ways in the three texts I discuss in this chapter, as they attempt to negotiate an experience of climate catastrophe that resists sublimation.

The texts I read in this chapter engage with the Burkean sublime on different levels. Byron's "Darkness", which I will discuss in the first part of this chapter, engages with the complexity of imagining an apocalyptic subject position without recourse

21 see Morton, *Hyperobjects*, 4. For a more detailed discussion of Romantic conceptualizations of "nature" including Morton's view see Rigby, "Discourses of Nature", 73–78.

22 Morton, 72.

23 Morton, 5.

to the numinous or a possibility of sublimation, but ultimately reinstates an external subject position through the construction of a dream framework that provides the poetic subject with the necessary distance to envision apocalypse.

"Darkness", and Mary Ann Browne's "A World Without Water", which I discuss in the second part of this chapter, are connected by this common device of the dream framework which they, however, use in different ways and for different purposes. The two texts also approach their respective apocalyptic events from slightly different angles which, as I will show in the following, involve the use of distinct poetic strategies of distancing and embodied experiences to support fundamentally different, ultimately gendered reactions to climate catastrophe.

As a substantially longer narrative text that draws on prophecy rather than dream as a framework of its futuristic scenario, Mary Shelley's *The Last Man*, which I discuss in the third part of this chapter, further complicates these arguments. As in the two poems, the novel models different, frequently gendered responses to (climate) catastrophe, raising questions about "nature", humanity and perception in the process but also reflecting Romantic poetics and aesthetics on a meta-narrative level. The novel's complex narrative structure, which I will discuss in more detail below, further complicates the question of situating knowledge and of positioning the perceiving and speaking subject in the context of (climate) catastrophe. The novel's narrative framework highlights central concerns with failed strategies of managing and critically framing different crises and the unsustainability of isolation, in both social and poetic contexts, but most prominently, in response to a catastrophe of apocalyptic proportions. Beer's argument about the contemporary critical tendency to frame apocalyptic texts within the current moment seems particularly appropriate for a novel such as *The Last Man* that predicts, in many ways, social and political reactions to a pandemic. My discussion here will, however, predominantly focus on reading the texts in the context of Romantic literary aesthetics and its framing of gender in the context of climate catastrophe, before I return to the discussion of their impact on current discourses of climate crisis in the final chapter on "Mushrooms" and my Conclusion.

George Gordon Lord Byron: "Darkness"

The winds were wither'd in the stagnant air,
And the clouds perish'd; Darkness had no need
Of aid from them – She was the Universe.[24]

As I argue in this section, "Darkness" engages with the problem of maintaining a stable poetic subject position in the middle of a secularized apocalypse that does not permit the possibility of religious or aesthetic sublimation.[25] Written during the summer of 1816 and included in *The Prisoner of Chillon* (published in the same year),[26] "Darkness" envisions a catastrophe of apocalyptic proportions, without being aware of the precise cause of the catastrophic events that framed its composition.[27] Line 16 and 17, "[h]appy were those who dwelt within the eye / Of the volcanos, and their mountain-torch,"[28] ironic in itself, because of the precarious position it describes, retrospectively adds an additional layer of irony to the poem's treatment of climate catastrophe, as anyone dwelling near, much less in, Mount Tambora in 1815 would not have survived its cataclysmic eruption.[29] The imagery evokes both the problem of positionality, as the speaker needs to be situated outside of the described events, and the Burkean sublime. The lack of distance between these ostensibly "happy" subjects and the observed "mountain torch" draws attention to the poem's complicated relationship with the aesthetics of the sublime, which it ultimately attempts to subvert, as I will argue below. The line's phrasing, beginning with the somewhat incongruous "[h]appy" which interrupts the regular iambic meter, suggests this is a relative happiness in a dire situation, as those who dwell in the eye of an active volcano are invariably also doomed, even if they are given a brief respite to allow their momentary enjoyment of this dangerous source of light in the utter darkness.

The poem's central apocalyptic scene of utter darkness is introduced within the first six lines:

---

24 Byron, "Darkness," 717.

25 see Phillips, "Metaphor and the Unprecedented: Byron's 'Darkness' and Responding to Ecological Disaster." James Phillips also reads the poem as an expression of aesthetic breakdown in the face of ecological disaster, focusing, however, on metaphor instead of positionality.

26 For an account of the poem's reception/popularity see Dingley, "'I Had a Dream...' Byron's 'Darkness.'"

27 Bate suggests that Byron's apocalyptic vision was likely triggered by the unusual number of sunspots (see Bate, "Living with the Weather.").

28 Byron, "Darkness," 716.

29 See Behringer, "Die Tamborakrise," 9. Wolfgang Behringer reports an estimated number of over 100.000 deaths in Indonesia alone, suggesting that several million deaths worldwide should be added to the immediate death toll of the volcanic eruption, as hunger and disease caused by crop failures spread across the globe.

The bright sun was extinguish'd, and the stars
Did wander darkling in the eternal space,
Rayless, and pathless, and the icy earth
Swung blind and blackening in the moonless air;
Morn came and went – and came, and brought no day,[30]

While the poem gives no reason for the original apocalyptic event, its past tense construction suggests an inevitability that is driven forward by the progressive forms of "darkling" and "blackening". The sixth line introduces a temporal framework of continuity, as at least two mornings pass without the arrival of daylight. The fact that the poem uses the word "[m]orn" to name the time of day is interesting in this context as both "morn" and "day" are, by definition, associated with the earth's position towards the sun and such concepts as "sunrise" and "daylight". "Morn" and "day", thus, become contested or even meaningless terms for a planet that does no longer experience sunlight, drawing attention to the ways in which a confrontation with an apocalyptic scenario of these proportions undermines fundamental cultural tenets such as time. The seemingly untethered use of temporality also draws attention to the speaker's position outside of the events, but the poem does not address the question of who is marking time in a dark universe, instead using concepts such as "morn" or "a moment" in conventional ways, before dissolving time along with human society and everything else in the universal darkness.

Along with the light, warmth is also taken away within the first lines of the poem, as the "icy earth / Swung blind and blackening in the moonless air". The cold is, in this case, caused by the absence of warming sunlight, involving more than one sensual perception in the experience. The spatial framework of the poem is planetary from the beginning, moving beyond the icy earth to include the sun, moon and stars, even if later in the text the fate of countries, groups of people and even individual humans and animals will be introduced as exemplary of humanity's hopelessness. As the poem is predominantly concerned with the fate of humanity as a whole, individual appearances are used to draw attention to the ways in which the apocalyptic events destroy the social fabric of human relations, the relationship between humans and animals and the relationship between humans and non-human nature.

The poem discusses the impact of the catastrophe on wild animals by describing them as behaving out of character – "the wildest brutes / Came tame and tremulous" –, or framing them in biblical language, which describes that even "vipers" sought the company of "the multitude" in line 35–37, although "[h]issing, but stingless – they were slain for food."[31] Any connection between humans and animals – let alone any other element of the non-human natural world – is, thus, denied early on

30 Byron, "Darkness," 716.

31 Byron, 716.

in the text as even domesticated animals are disconnected from their human masters. The second longer episode in the poem concerns a single dog that doesn't attack its master but "was faithful to a corse, and kept / the birds and beasts and famish'd men at bay" only to die "with a piteous and perpetual moan, and a quick desolate cry, licking the hand / Which answer'd not with a caress" (lines 49–54).[32] While faithful dogs have become a staple of post-apocalyptic fiction and film in the twenty-first century, it is interesting to note here that the dog in its germinal appearance is also a vehicle of irony, suggesting the futility of the bond between the human master and his faithful canine companion.[33] The nuclear family – another staple of post-apocalyptic texts – does not appear at all, underlining the overall impression of the complete breakdown of the social fabric: "no love was left; / All earth was but one thought – and that was death / Immediate and inglorious" (lines 41–43).[34] Human and non-human life emerges as equally helpless and pointless, as the poem's final, universal darkness swallows every kind of meaning beyond its own presence.

Biblical Language and the Secular Apocalypse

The biblical serpent has lost its literal and metaphorical sting, as humanity can no longer be tempted into doing anything but focus on bare life. The fight for survival is Hobbesian in its brutality as "a meal was bought / With blood and each sate sullenly apart / Gorging himself in the gloom: no love was left / All earth was but one thought – and that was death."[35]

While humanity's "selfish prayer for light"[36] in line 9 seems to evoke a religious framing, there is no sign of divine intervention – neither in the cause of the apocalypse, nor in its process, nor beyond its end. On the contrary, the dark universe is decidedly empty of anything, and religious thought stops with the end of humanity – and indeed all life – in line 69–71: "The world was void, / The populous and the powerful was a lump, / Seasonless, herbless, treeless, manless, lifeless – / A lump of death – a chaos of hard clay."[37] The void is not filled with a sense of the numinous, as all meaning – even the biblical language of creation – disappears with the degeneration of "nature" into "a chaos of hard clay".

"Darkness" can be read as a reversal of biblical genesis, which begins on an earth populated by many species and, after describing their extinction, ends in a "void". As Morton D. Paley argues, "[w]hen in certain poems Byron makes apocalypse his

32 Byron, 717.

33 Nandita Batra presents an argument against this anthropocentric focus on animals in Romantic literature. (see Batra, "Dominion, Empathy and Symbiosis.").

34 Byron, "Darkness," 717.

35 Byron, 717.

36 Byron, 716.

37 Byron, 717.

central subject, it is unaccompanied by millennium."[38] Paley's argument centers on the secularization of apocalyptic imagery displayed by some of the younger Romantics (among them Byron and Mary Shelley) – a tendency which distinguishes their work from earlier Romantic apocalyptic visions by William Blake and William Wordsworth.[39] These texts are, to various extents, concerned with "lastness"[40] and the revelation of last things, but, so far as they draw on Byron as a model, "[m]illennial associations are consistently invoked in order to be bitterly frustrated".[41] As Paley shows, however, by comparing Byron's journal entries concerning his travels in the Alps with his poetic vision of Apocalypse in *Manfred*, apocalyptic imagery in Byron's work is also closely tied to the observation of natural phenomena,[42] inviting ecocritical readings.

Situated in the early 2000s, Paley's framing of "Darkness" as an apocalyptic text reflecting biblical rather than ecological images shows that the shift in critical interest towards climate crisis as a major theme of literary and cultural production is a relatively recent one. It also draws attention to the ways in which Romanthropocene criticism has begun to reframe the Romantics in the light of anthropocene discourses. This is a gradual process, which moves from Bate's ecocritical approach (in 1996), arguing that "'Darkness' begins with the stress of Weather"[43] to reframe the poem in the light of the Tambora crisis as a "global struggle for subsistence [which] leads ultimately to the extinction of mankind",[44] towards more recent interpretations of "Darkness" as an early anthropocene text,[45] which can be interrogated with regard to its understanding of climate catastrophe in the early anthropocene.

In both Byron's and Mary Shelley's work, which I will discuss below, apocalypse has been uncoupled from its earlier meaning of Christian revelation in the sense that the "last things" emerging in both texts revolve around the largely secularized materiality of existence on earth (and of earth) instead of the sacred mysteries of life hereafter. Both texts still convey a sense of awe on the part of the speaker or narrator, but the objects that trigger this overwhelming sense of greatness are perceived as firmly residing in the realm of the natural and not the supernatural world, further complicating attempts at aesthetic engagements with these objects. This also moves the texts closer to a more contemporary understanding of climate catastrophe as neither humanly controllable nor subject to divine intervention, which might suggest at least one reason for their appeal to Romanthropocene criticism.

---

38 Paley, "Byron," 3.

39 For further discussion of these see Paley, *Apocalypse and Millennium in English Romantic Poetry*.

40 Paley, "Byron," 4.

41 Paley, 8.

42 Paley, 1.

43 Bate, "Living with the Weather," 433.

44 Bate, 434.

45 see Reno, *Early Anthropocene Literature in Britain – 1750–1884*.

The Impossibility of Apocalyptic Sublimation

The Romantics' aesthetic framing of awe and terror draws predominantly on Edmund Burke, as I have argued above. For Burke aesthetic experiences are, however, not only ultimately rooted in divine creation, for him the God of biblical scripture also serves as a pivotal example of unfathomable greatness and terror.[46] To imagine the end of humanity or, in Byron's case, the solar system, without any glimpse of the divine reduces the aesthetic distance that is a necessary component of the experience of the sublime, as it delivers the human observer to a merciless universe without a possibility of salvation or redemption in another life. The experiences associated with the end of the world remain ungraspable to the rational mind, but the aesthetic pleasure associated with them can no longer be derived from "the removal of pain",[47] because there is no sense of "escaping imminent danger"[48] in this kind of secularized apocalyptic scenario. The sense of an ungraspable vastness conveyed in "Darkness" complicates the aesthetic category of the sublime by taking it out of the context of creation and offering no possibility of sublimation through the experience of anything beyond the text. The absence of sacred or numinous meaning within the apocalypse thus complicates the texts' relationship to the sublime.

By ending in universal darkness, Byron's poem also discards the sublime into the dark void at the end of the poem – alongside other affective sensations. The text allows the perceiving subject no sublimating distance to the perceived sublime object and aesthetic experience is, thus, annihilated along with everything else. In the absolute absence of light "men forgot their passions in the dread / of this their desolation; and all hearts / Were chill'd into a selfish prayer for light."[49] While the prayer for light seems to resonate with Burke's definition of darkness as terrible, the poem's phrasing already suggests that the terrible darkness has come too close and does no longer allow for any sort of emotional reaction (or "passion"), nor does the poem convey a sense that their "selfish" prayers might be answered. In a manner that resembles the poem's treatment of biblical language, the sublime is, thus, also evoked as a possible aesthetic register that does, however, not hold up to the test of utter hopelessness in the face of global apocalypse. The poem's structure supports this reading as the notion of privation is evoked over and over again in a number of words ending in "-less" and highlighted by an interruption of the regular iambic meter in line 71, which lists several adjectives ending in -less: "Seasonless, herbless, treeless, manless, lifeless".[50] While the text is written in blank verse with very few deviations,

---

46 see Burke, *A Philosophical Enquiry into the Origin of Our Ideas of the Sublime and the Beautiful.*

47 Burke, 31.

48 Burke, 32.

49 Byron, "Darkness," 716.

50 Byron, 717.

these irregularities occur predominantly at the beginning of the line, where the first syllable of certain words is stressed, resulting in a dactylic pattern, before the line returns to the iambic meter of blank verse. These stresses often support ironic constructions that are challenged by the remainder of the line, such as "happy" referring to the reckless people who have fled into an active volcano. These ironic constructions challenge the emergence of stable meaning, thus underlining the poems resistance to sublimation.

For Burke the sublime is frequently associated with a form of "privation",[51] that is also a dominant theme in the poem. While Burke provides a physical explanation of the pain caused by darkness that is based on the contractions of the iris,[52] he also associates darkness with the absence of light and sight and the fears associated with this absence, as one of the forms of privation that evoke terror,

> for in utter darkness, it is impossible to know in what degree of safety we stand; we are ignorant of the objects that surround us; we may every moment strike against some dangerous obstruction; we may fall down a precipice the first step we take; and if an enemy approach, we know not in what quarter to defend ourselves; in such a case strength is no sure protection; wisdom can only act by guess; the boldest are staggered, and he who would pray for nothing else towards his defence, is forced to pray for light.[53]

There are several resonances between "Darkness" and Burke's explanation of darkness as a cause of the sublime (from the prayer for light to the notion of encountering an enemy in the dark, which I will discuss below), which suggest a deliberate connection between the two texts that would support the impression of an ironic engagement with sublime aesthetic. Burke's explanation of why darkness "is terrible" functions along the same lines as the other forms of privation and sources of the sublime he examines – vacuity, solitude and silence – namely by presenting them as "terrible" based on personal observation, anecdote and literary quotations.[54] For Burke, distance between the observer and the observed sublime object is a key element in all of these cases, as a lack of distance would result in immediate danger instead of aesthetic pleasure. As Geoff Payne shows, a prerequisite for the experience of the sublime is "a rigid division between subject and object [...] so that the security of the self can be guaranteed against the dangers contained in the sublime object."[55] This clear division of subject and object has vanished in "Darkness", as the

51 Burke, *A Philosophical Enquiry into the Origin of Our Ideas of the Sublime and the Beautiful*, 65.

52 see Burke, 132–33.

53 Burke, 130.

54 see Burke, 65.

55 Payne, *Dark Imaginings*, 41.

experiencing and the speaking subject dissolves in the universal darkness, thus undermining the possibility of aesthetic experience. The other texts I will discuss in this chapter deviate from this model in distinct, frequently gendered ways, by complicating the experience of the sublime through the reduction of distance between the experiencing subject and the sublime object – for instance by placing the speaker as an experiencing subject in the middle of the apocalyptic scenario.

"Darkness", however, negotiates this problem in an almost literal manner, challenging Burke's theory, as the feminized darkness of the universe swallows everything at the end of the poem, negating the possibility of distance, agency and sublimation. The regularity of the blank verse and the general structure of parallelisms and enumerations results in a pace that seems to be moving forward relentlessly towards this seemingly inevitable ending. Most lines end on a stressed syllable, with the only exception to this pattern occurring, again, in line 71 – the only line that is merely an enumeration of adjectives, all ending in "-less" – which underlines its importance for the text. Along with "Rayless, and pathless" (used to describe the stars' wandering "in the eternal space" in line 3), "moonless" (line 5), the "useless wings" of birds (in line 34), and the "stingless" vipers (line 37) as well as the "tombless" bones of men (line 45) and "sailorless" ships (line 75), these adjectives seem to stress a non-Burkean aesthetic of privation, as more and more elements of "nature" and finally also human beings are swallowed by the darkness, ending in "lifeless" as a general summary of the poem. This connection between "darkness" and the adjectives ending in "-less" is underlined by the fact that their repetition creates one of the few rhyming structures in the otherwise rhymeless poem. This pattern might be extended to include "hideousness" (line 67) and, perhaps to a lesser extent, "mistress" – referring to the moon (line 79). "Hideousness" evokes another almost mocking parallel to Burke's explanation for why darkness is sublime – the scenario of encountering an enemy in the dark (quoted above). The word occurs as the conclusion to one of the longer individual anecdotes in the poem (line 55–69):

> [...] but two
> Of an enormous city did survive,
> And they were enemies: they met beside
> The dying embers of an altar-place
> Where had been heap'd a mass of holy things
> For an unholy usage; they rak'd up,
> And shivering scrap'd with their cold skeleton hands
> The feeble ashes, and their feeble breath
> Blew for a little life, and made a flame
> Which was a mockery; then they lifted up
> Their eyes as it grew lighter, and beheld
> Each other's aspects – saw, and shriek'd, and died –

Even of their mutual hideousness they died,
Unknowing who he was upon whose brow
Famine had written Fiend.[56]

The passage evokes the idea of organized religion in the form of the "altar place", that, interestingly, remains as non-denominational as do the "holy things" that have been "heap'd [...] / For an unholy usage". The assumption of their holiness is based on their history, not their current situation as a "mass" (perhaps a pun on the Catholic mass, that they will no longer be part of) that has either been piled up for looting or to be set on fire – the poem is not clear on whether it is the pile of "holy things" itself or the church that has now been reduced to "dying embers", but both have been removed from their original context. The passage as a whole conveys irony on different levels: first, all the "feeble" raking, scraping and blowing of breath results in the much-needed fire, which leads to a brief lifting of the complete darkness that allows the dying enemies to behold "[e]ach other's aspects" only so they can – and this is the second layer of irony – die "of their mutual hideousness". While both parties are presumably almost "famished" along with everyone else and scrape together the "feeble ashes" with "cold skeleton hands", they do not die of hunger, but of "hideousness" and "[u]nknowing who he was upon whose brow / Famine had written Fiend". The alliteration and capitalization of "Famine" and "Fiend" emphasizes the connection between hunger and ugliness (or "hideousness" as the poem puts it), turning emaciation from a bodily state into an aesthetic category – and one that carries deadly strength. Famine also presents one of the connecting concepts that allow for a reading of the poem in the light of climate catastrophe and the chain of climate events triggered by the eruption of Mount Tambora. Read in the context of the Burkean sublime, this episode conveys the necessity of distance from the terrible object, as the enemies are unable to sublimate their experience into a form of delight and, instead, die of another aesthetic experience. As such, their death of "hideousness" can also be read as a form of meta-aesthetic commentary on the uselessness of the sublime as an aesthetic category in the absence of a perceiving subject with a sufficient distance to the object of terror. The passage carries an additional suggestion of lastness, as the two are the last "[o]f an enormous city" and after they die (the word "died" is interestingly used twice to underline their mirrored deaths), humanity only appears in the poem as an absence ("manless" in line 71 and "sailorless" in line 75).[57]

56 Byron, "Darkness," 717.

57 Byron, 717. In this sense, "Darkness" can not really be read within the context of the 'last man'-theme, as there really is no last man and no post-apocalyptic world to inhabit (see Schroeder, "Byron's 'Darkness' and the Romantic Dis-Spiriting of Nature," 117.).

Gendering the Apocalyptic

Gendered assumptions around the sublime also emerge in interesting ways in "Darkness" as darkness – the ultimate privation of light – is not only feminized but also taken out of the context of male agency, contesting the binary structure of the sublime and the beautiful in Burke's aesthetic that I have outlined above. The poem mainly uses male gender markers to describe the universal experience of devastation throughout: "men forgot their passions in the dread" (line 7), "men gather'd round their blazing homes" (line 14) and "the brows of men by the despairing light / Wore an unearthly aspect" (lines 22–3) as well as "men / Died" (lines 44–5).[58] While "man" is, of course, used as a universal term in the sense of "humans" and there are a number of gender-neutral groupings that would presumably have been meant to include women and children, e.g. "those who dwelt within the eye / Of the volcanos" (lines 16–17), "the meagre by the meagre were devour'd"[59] and "the crowd was famish'd by degrees" (line 55),[60] it is interesting that the poem does not distinguish male and female reactions and generally tends to focus on the actions and reactions of "men", which range from prayer, weeping, teeth-gnashing and despair to war and destruction. Prayer for light is the earliest recourse, appearing already in line 9, not to be repeated throughout, even as the situation becomes more dire. Hope – already framed as "fearful" – makes a brief appearance in line 18, never to be seen again and "[w]ar, which for a moment was no more / Did glut himself again" (lines 47–48).[61] Any hint at a social fabric – and there are few to begin with – has collapsed early on as (in lines 11–14):

The palaces of crowned kings – the huts,
The habitations of all things which dwell,
Were burnt for beacons; cities were consum'd,
And men were gather'd round their blazing homes.[62]

This attempt at creating a last bit of light by burning their homes emerges as another ironic construction in the text, as it destroys the possibility of gathering in any form of supportive community and creates more (smoky) darkness in the process.

Payne argues that the "old world that crumbles under the weight of such darkness is an essentially masculine domain, its landscape dominated by identifiable

58 Byron, "Darkness," 716–17.
59 Byron, 716.
60 Byron, 717.
61 Byron, 717.
62 Byron, 717.

symbols of the patriarchal order (castles, temples cities)."[63] This destruction is, however, largely self-inflicted through what Payne describes as "masculine modes of thought"[64] which emerge in the destructiveness and futility of responses to catastrophe. Beyond the described self-annihilation of human culture, the poem's description of humanity's devastation of the livable environment in the face of catastrophe is extended to the planet as a habitat, as "[f]orests were set on fire", once again granting only a brief period of light as "hour by hour / They fell and faded – and the crackling trunks / Extinguish'd with a crash – and all was black" (lines 19–22).[65] The poem's presentation of these destructive patterns as the only response to climate catastrophe of apocalyptic proportions makes it possible to reframe the text in the light of recent discussions of gendered responses to climate crisis and the adherence to fossil fuel economies (which I will discuss further in my Conclusion), but it also suggests a reading of "Darkness" in conjunction with other early anthropocene texts that imagine gendered responses to apocalyptic scenarios.

While most of the reactions to the utter desolation are presented with either a male or a gender-neutral framing, female pronouns, interestingly, appear only at the end of the text, after the extinction of humanity. There are two female instances – the moon, introduced as the already absent "mistress" of the tides (in line 79) and "Darkness" herself in the poem's last line. While the female moon is not an unconventional construction, especially in the context of Romance languages, the female connoted and capitalized "Darkness" that ends the poem by encompassing the Universe requires a closer analysis as it serves as the poem's agent of destruction and underlines the ultimate denial of a sublime aesthetization of catastrophe.[66]

The last line, in which this image occurs, is broken in half by a caesura, with the first half finishing the previous line's sentence "Darkness had no need / Of aid from them"[67] (with "them" referring to the winds and clouds). This is one of only a few caesuras in the poem which mark the sense of an ending – the desolation of mankind in line 8 and the death of the two enemies, the last survivors of a city, who die of "mutual hideousness" in line 69 – before the dominance of the feminized and capitalized "Darkness" marks the ultimate ending. Darkness as the subject of the last line has already been introduced in the previous line, setting the stage for the use of

63 Payne, *Dark Imaginings*, 34.

64 Payne, 34.

65 Byron, "Darkness," 716. Julian S. Whitney argues that "Lord Byron uses multiple images of consumption in 'Darkness' to engage in a broader conversation about the merits of conservation and resource management." (Whitney, "Planetary Crisis," 35.).

66 Payne shows that most readings of the poem's gender constellation reach for biographical explanations, seeing "Darkness" as an outlet for Byron's frustration with his exile and failed marriage – with the exception of Cynthia Burkhead's reading, which focuses on the poem's general expression of misogyny (see Payne, *Dark Imaginings*, 32–33.).

67 Byron, "Darkness," 717.

the capitalized female pronoun after the caesura. The phrase "She was the Universe" suggests a full sentence, but the line is broken in half by a hyphen instead of a full stop, connecting the last few words to the previous line. While the phrase seems to hint at a metaphorical construction with the Universe as the vehicle, the universality of the claim stands in the way of a distinct interpretation of the tenor beyond the straightforward suggestion of universality, further thwarting the possibility of sublimation. Along with everything else, meaning and the possibility of interpretation are also lost in the universal darkness at the end of the text.

The capitalization and gendering of "Darkness" as "She" who "was the Universe" further complicates the relationship of experiencing subject and experienced, ungraspable object in the poem. In "Darkness" the female is clearly neither associated with the beautiful – as Burke suggests – nor with the sublime, which is eliminated along with the possibility of sublimation. Instead, the female Darkness who "was the Universe" is constructed as the ultimate privation – the absence of light that causes a darkness so complete that it has "no need / Of aid from" the "clouds" and "winds [that] were wither'd in the stagnant air" or any other dark or darkening element of "nature" to become universal.[68] All connections to "nature" – or what little remains of it – are severed by line 81, as "Darkness had no need / Of aid from them". She is at the same time universal and self-contained, a universal power of nothingness that also ends the poem, as the end of the creative imagination coincides with the end of "nature". This has important poetic ramifications: the poem's ending suggests a reading of creativity and destructiveness as gendered opposites, as male-connoted creativity is pitted against female-connoted deprivation/nothingness. This almost triumphant proclamation of the power of universal darkness evokes the aesthetic strategies of the sublime without the possibility of sublimation in the absence of an experiencing subject situated at a suitable distance from the catastrophe described in the text. The entirely uninvolved speaker's situational and emotional distance from the narrated events similarly negates the possibility of reading them within the aesthetic framework of the sublime.

The fact that "Darkness" is gendered as female at the end of the poem has led to a number of readings focusing on the triumph of an all-encompassing female figure over male-encoded power structures that are deconstructed throughout the text. While this would be a compelling reading for a text interested in undermining strict, gendered dichotomies (male/female, subject/object, nature/culture, light/darkness)

68 Byron, 717. Payne suggests a figurative reading of these verses in the context of Byron's frustration with "the 'conspiracy' of women" supporting his wife in their separation, which he saw as instrumental in ending his marriage. While such a reading is entirely possible, I would agree with Payne, that "the poem invites a far greater range of interpretive possibilities" (Payne, *Dark Imaginings*, 34.).

in Romantic aesthetics, Byron's focus ultimately lies elsewhere – namely on challenging Burkean sublime aesthetics as a whole, without undermining the possibility of poetic authority. While the poem, thus, ends in a feminized darkness, there are, after all, two (gendered) pronouns bracketing the poem's structure: the opening "I" of the dreamer and the "She" referring to the darkness who has become the universe in the last line. The use of the female pronoun for and capitalization of "She", "Darkness" and "Universe" is fraught with meanings for the poem's reverse creation myth, as it imbues the eponymous "Darkness" with all-consuming power, while at the same time depriving it of all connotations of potential becoming and creative agency. In a reverse act of creation, the speaker's position disappears behind the descriptions of the devastation wrought in the name of darkness. But, while the poem's gendered focus moves from a presumably male creator (in this case in the sense of the poetic creator of apocalyptic imagery, not of a creator god of the world itself) to a female void swallowing everything in complete darkness, the subject position and narrative authority remain with the distant, disembodied speaking voice.

Reinstating a Subject Position

The (de-)construction of the world by a disembodied voice that "Darkness" envisions relies on an understanding of poetics that envisions the poet as a figure who possesses the fundamental ability of creating truths beyond the necessity of physical experience of and embodied representation in the material or the textual world. In this understanding, the fact that "Darkness" does not situate the speaker's voice within the described scenario and, consequently, does not envision an "end" to the speaker's existence alongside everything else, becomes a powerful poetic strategy that allows the speaking subject to reclaim authority over the way the events are framed within the poem.

The speaker conjures the apocalyptic vision under the guise of the ominous semi-reality status of "a dream that was not all a dream."[69] In the Romantic context, this can not necessarily be interpreted within a psychoanalytic framework. As Anita O'Connell argues, in the Romantic period, daydreaming, dreaming and theory of mind were considered "a spectrum of liminal imaginative states"[70] and, thus, not only connect the literary imagination to more closely imaginative dream states than post-Freudian psychology's interpretation of dreams as products of the unconscious mind, but also confer a certain creative or poetic power to the dream imagination. As O'Connell argues, "[i]n the Romantic period the creative vision in the mind could refer to the inner experience of any of a range of imaginative states, from thoughts and ideas to the vision that is imagined when reading, composing

69 Byron, "Darkness," 716.

70 O'Connell, "Visions in Verse: Writing the Visual in Romantic Dream Visions," 36.

poetry, daydreaming, in reverie, madness, or sleeping dreams."[71] The Romantics draw, as O'Connell points out, on a tradition of Cartesian thought that did not only see the imagination as the "image making facility" but also thought ideas "to be essentially visual, and thus involved with the imagination."[72] Dream visions, in this sense, possess a prophetic quality – not in that they *predict* the future but in that they *create* as close as possible an image of an idea. The images in dream visions are, thus, imbued with a reality status beyond the expression of the poet's unconscious mind, as they emerge as poetic expressions of ideas, and "Romantic dream visions are often concerned with the disparity between the vision in the mind – the poet's original conception – and its linguistic expression in poetry".[73] Byron's "dream that was not all a dream"[74], thus, emerges as a meta-poetic comment on the poet's ability to aesthetically frame this vision of apocalypse. The opening dream framing allows Byron to reintroduce a poetic authority-figure in command of but residing outside of the apocalyptic world of the text. It is in this sense that the speaker/dreamer is able to envision the end of everything in universal darkness without shifting the authority over this narrative to darkness at the end of the text.

As I have argued above, the figuration of the speaker as absent from the text removes the poem from the aesthetic framework of the sublime, which hinges on the pre-rational confrontation of the perceiving subject with an object of power to be resolved in a moment of aesthetic delight. Payne also suggests that the poem evokes sublime imagery, which does, however, not fulfill the sublime effect it promises, thus causing uneasiness among reviewers who attempted to frame the poem within the contemporary aesthetic discourse of the sublime and the beautiful.[75] This uneasiness draws attention to the ways in which shifting discourses around the possibility of representing the unimaginable emerge in Romantic poetics, but the speaker's unresolved, floating position as the poetic initiator of this dream vision, who remains distanced from the apocalyptic events, also draws attention to a problem of positionality that resonates with anthropocene contextualizations of human positionality in the face of climate crisis.

It is possible to read "Darkness" as a text about climate change, focusing, as so many climate narratives, on humanity's efforts to survive the apocalypse by any means possible.[76] Such a straightforward thematic reading in light of a more recent

---

71 O'Connell, 35.

72 O'Connell, 37.

73 O'Connell, 40.

74 Byron, "Darkness," 716.

75 see Payne, *Dark Imaginings*, 52–53.

76 A contemporary example of this kind of anthropocentric climate narrative is Roland Emmerich's 2004 action film *The Day After Tomorrow*, which shows the last survivors of a devastating climate catastrophe freezing the northern hemisphere as they burn rare books in the New York Public Library to keep warm (see *The Day After Tomorrow*.).

understanding of climate crisis can, however, not entirely account for the poem's structural peculiarities – most prominently its dream framework and its focus on the annihilation of sublime aesthetics in the context of apocalypse. I will turn to the question of how dreams can function as a more situated and less authoritative framing device in my discussion of Mary Ann Browne's poem below, reading "A World Without Water" as an example of an early anthropocene text which engages with the exigencies of climate crisis as a crisis of humanity as well as aesthetic representation. As I argue Browne uses the dream framework of her poem to frame and construct an embodied, situated subject position and experience that allow her to explore an apocalyptic scenario from a non-authoritative perspective, negating the necessity of an experiencing autonomous self and presenting connection and compassion as central elements in the context of climate catastrophe.

Mary Ann Browne: "A World Without Water"

While "Darkness" is often read as Byron's poetic reaction to the effects of "volcano weather"[77] and sunspots in the summer of 1816,[78] much less is known about the circumstances that triggered the composition of "A World Without Water". Its composition date is uncertain, but might lie between 1829 and 1832, as Mary Ann Browne was born in 1812 and published a first volume of poetry, *Mont Blanc and Other Poems by Mary Ann Browne, in her Fifteenth Year* in 1827, followed by two other volumes in 1828 and 1829 before the publication of "A World Without Water" in the *Winter's Wreath* in 1832. While the devastating impact of volcano weather following the eruption of Mount Tambora in 1815 is relatively well documented, climate records covering precipitation and droughts in the early nineteenth century are harder to come by. Murphy et al. have reconstructed climate data for the period of 1748 to 2000, observing two major periods of drought in the British isles between 1765 and 1768 and between 1834 and 1836, but also citing earlier work which observed more localized drought phenomena.[79] None of these dates coincide exactly with the probable composition date of Browne's work, but the scarcity of weather data and the lack of biographical knowledge about Browne make it impossible to rule out personal experience of drought as an inspiration for the poem.

As "Darkness", "A World Without Water" is framed as a dream, which takes the idea of climate catastrophe to an imagined extreme that moves far beyond the possibility of personal experience, thus resisting a mimetic interpretational framework.

---

77 see Stommel and Stommel, *Volcano Weather*.

78 see Higgins, *British Romanticism, Climate Change and the Anthropocene – Writing Tambora*.

79 see Murphy et al., "The Forgotten Drought of 1765–1768: Reconstructing and Re-Evaluating Historical Droughts in the British and Irish Isles."

Dissimilar to "Darkness", this dream is presented as a complete framing device, allowing the speaker to emerge from the vision, while also carrying over a residual sense of the imagined "agony". Where Byron's speaker remains, as I have argued above, distanced from the apocalyptic scenario in his bid to challenge sublime aesthetics without ceding control of the authoritative voice of the poetic subject, Browne's poem pursues a poetic strategy of situated involvement, which allows her to manifest a poetic voice without claiming authority over the situation or its aesthetic framing. These distinct strategies reflect differences in gendered poetics as well as the problem of aesthetically framing apocalyptic events in an increasingly secularized world, which might be overlooked behind the more obvious thematic similarities between the two texts: both imagine inescapable apocalyptic scenarios without a redemptive religious component and both frame them as dreams, thus positioning their overt speakers at a certain experiential distance from the fatal reality of the apocalyptic events. Both poems also foreground the effects of and human responses to climate catastrophe, using the experience of individual but nameless humans as representative of humanity in general. The possibility of an intertextual reading of Browne's poem in the light of Byron's is further supported by Browne's open acknowledgement of another influence from the Romantic canon: the poem is not only preceded by an epigraph from Coleridge's "The Pains of Sleep" (written in 1803, but not published until 1816), but it is also formally indebted to Coleridge's work – for instance in its use of the ballad stanza with certain deviations that I will discuss below.

Structuring Scarcity

"A World Without Water" is substantially longer and less evenly paced than "Darkness", extending its apocalyptic narrative to sixteen stanzas of varying lengths, that consist of 6 to 12 lines, with 8 stanzas of 8 lines and only one – stanza 7 – using an uneven number of 7 lines.[80] This curiously symmetric asymmetry at the center of the text supports the effect of suspense created at the end of stanza 7, which ends on the unfulfilled hope for "some blessed dew" that might arrive on the night wind to end the drought. This hope is dashed in the first line of stanza 8, which opens on the words "Vain was the hope!"[81] The suspense is further supported by the rhyme scheme, which varies between various cross-rhyme patterns throughout the poem, but changes in the last two lines of stanza 7, which rhyme with each other. This structure creates a pattern of suspense, which suggests a dramatic reading or performance before an emotionally engaged audience rather than a solitary reader.

80 This would be too much of a coincidence to not have some numerological meaning, which I will, however, not delve further into here.

81 Browne, "A World Without Water," 156.

Unlike Byron's "Darkness" which is composed in blank verse with only a few deviations from the regular iambic meter (as I have shown above), Browne's "A World Without Water" frequently adds unstressed feet to the generally iambic ballad stanzas. There are two recognizable patterns for these added syllables: the first is an unstressed "and" added at the beginning of some lines – a pattern not uncommon in performative (oral or musical) ballad traditions, which underlines a sense of spontaneity and emotional involvement on the part of the speaker or singer by creating a suspenseful pause through the addition of a second unstressed syllable before the first stress in the line. The second is an addition of at least one, sometimes several unstressed syllables throughout a line which results in a dactylic rhythm that interrupts the iambic flow of the line – for instance in line 6: "I *thought* there had *fa*llen no *coo*ling *rain*".[82] These dactylic additions occur most frequently in lines or stanzas that mention sources of water (the rain in stanza 1, a spring in stanza 3, water and the sea in stanzas 12 and 13, for instance), so one possible interpretation might be that the meter supports the poem's main contrast between the complete stillness of the environment without water (and also without wind or rain) with the unpredictable, skipping movement of bodies of water, supporting the latter with a meter that skips along somewhat unpredictably, adding unstressed syllables seemingly at random.[83] This reading might be supported by the unequal length of the stanzas, which create a rolling wave-like pattern in a poem that dedicates four stanzas to the description of the waterless sea alone. While water is thus constructed as an absence throughout the text, the poem reintroduces a memory of its movements through the use of formal elements.

The poem's alliterating title, with the abundance of "w"s again recreating a wave pattern, draws attention to the general theme of deprivation – of what is not there in the world described by the speaker, but everywhere in the formal elements and imagery of the poem. As "Darkness", "A World Without Water" does not construct this deprivation within an aesthetic framework of the Burkean sublime, but instead relies on a multitude of formal and perceptual details to create an overwhelming sense of urgency and need. Dissimilar to "Darkness", Browne's speaker is emotionally involved with this need, constructing further parallels of suffering – both human and non-human – throughout the poem.

These effects are supported by deviations from the regularity of iambic meter, as a few lines with dactylic additions do not refer to water in any shape or form but are instead concerned with other elements of the non-human natural world which are associated with the hope for shade, cooling temperatures or dew (for instance

---

82 Browne, 155 (my italics).

83 While it might be tempting to read these additions as a certain clumsiness with meter, it is highly unlikely that Browne, who also wrote and published musical scores, was completely unaware of the skipping rhythmical pattern she was creating.

trees, storms, the sky and night). Death, even in metaphorical use, also prompts the addition of extra dactylic syllables for emphasis, for instance, "in the dead of night" in line 1 and "[a]s though *Death* were too *busy* with other *things*" in line 33, supporting the personification of Death and the extended discussion of dry corpses in stanza 6 – which I will return to further below.[84]

Envisioning Embodied Experience

By emotionally highlighting the effects of drought on the non-human natural world, Browne's nightmare of an environmental catastrophe adds an element of specificity to the vision that is absent from the dream model she intertextually draws on (Coleridge's "The Pains of Sleep"), as well as Byron's distanced nightmare vision discussed above. Coleridge's speaker is "confused" and unsure if he "suffered, or [he] did" the ominous "deeds" he is tortured by in his sleep, focusing on his emotional reactions of "[l]ife-stifling fear, soul-stifling shame."[85] Browne's poem, by comparison, focuses on a clear nightmare vision of prolonged drought and its impact on both humanity and the non-human natural world. The sense of guilt the speaker seems to feel in Coleridge's poem is absent from Browne's, but the poem's language and description of people's reactions to the prolonged absence of water creates an ominous undertone of martyrdom, supported by the complete inactivity of everyone involved. While there is no discussion of an anthropogenic cause of the acute climate crisis in the poem itself, the abundance of suffering without a possibility of redemption draws attention to the absence of a sense of divine providence on this doomed planet.

Where conflict dominates people's reactions in "Darkness", the main reactions of humanity described in Browne's poem are passive: seeking shade, hoping, crying or having visions. While, unlike in Byron's "Darkness", no one is shown to be praying for heavenly delivery, the poem offers some religious references in the speaker's reactions to the scenery of devastation. When the speaker turns "away from earth and sea, / and looked on the burning sky," in the penultimate stanza, they observe that "no drop fell, like an angel's tear – / The founts of Heaven were dry."[86] The fact that "sky" and "Heaven" are mentioned in the same stanza, but only "Heaven" is capitalized suggests not only a difference in meaning, but a non-literal interpretation of the "founts of Heaven" as the grace of a benevolent deity. This reading of another godforsaken world might be further supported by an intertextual connection to Charlotte Smith's poem *The Emigrants* which uses the phrase "World of Waters" as synonymous

84 Browne, "A World Without Water," 155–56 (my italics).

85 Coleridge, "Letter from S.T. Coleridge to Robert Southey, 11 September 1803 (including early version of 'The Pains of Sleep')," 513.

86 Browne, "A World Without Water," 158–59.

with God's creation.[87] In this sense, a "World Without Water" would then also be a world abandoned by a Christian god.[88] The fact that neither the speaker, nor the people described in the text experience a sense of sublimation also removes Browne's poem from the context of the Burkean sublime. Instead, the poem is focused on the situated experience of its climate vision, which resonates with an understanding of the connections between human beings and the natural world beyond their aestheticization. The poem's ecocritical undertones which suggest a reading of the prolonged drought as "nature"'s backlash rather than divine punishment may not have been quite as obvious to readers in 1832, but the poem's "dream of agony" still resonates with Coleridge's "anguish and agony" in the epigraph, suggesting a strong undertone of confused guilt and remorse on the part of the dreamer, which is, however, not transferred into a call to action.[89]

The apocalyptic scenarios envisioned in "Darkness" and "A World Without Water" seem to preclude any kind of action beyond the immediate moment, but where Byron's "Darkness" is full of biblical imagery, creating a dynamic reverse creation narrative, the dominant sentiment in Browne's nightmare vision is the agony of utter stillness. It is no coincidence that the nightmare that tortures the speaker comes "in the dead of night", as there is almost no movement in the poem, apart from the speaker's own reactions. The nightmare framework is introduced unambiguously as "[a] dream of agony" in which the speaker "thought the world stood in affright".[90] The verb "stood" seems a curious choice here, but it sets the tone for the atmosphere of stillness that runs through the rest of the poem. The speaker's position, which is distinctly situated "on a hill" allows them to "look[...] all around" without having to move around.[91] In this context, "still" in line 11 carries a double meaning as "still / Strength in my limbs was found" suggests not only a temporal component referring to the fact that "there had fallen no cooling rain / For months", but also to the literal stillness of the speaker's position overlooking the parched landscape.[92]

Along with this clearly defined location of the speaker in the world of the text, the reference to "limbs" draws attention to the fact that this is an embodied pres-

---

87 Smith, *The Emigrants*, 2. For a discussion of Smith's work, see Day, *Romanticism*, 36.

88 While Browne was generally known to be religious and published hymns and, among other volumes of poetry, *The Coronal; Original Poems Sacred and Miscellaneous* (London, 1833), the few references to "Heaven" throughout the poem and the absence of any references to God, prayer or other religious elements suggests that "A World Without Water" is not, in fact, meant to openly transport a religious meaning, beyond the absence of salvation. The problem of theodicy further complicates reference to divine creation or salvation in secular apocalyptic contexts.

89 Browne, "A World Without Water," 155.

90 Browne, 155.

91 Browne, 155.

92 Browne, 155.

ence, clearly different from the disembodied and distanced speaking voice in Byron's "Darkness". This deliberate situating of the speaker within the physical world of the text does not only create a sense of connection to the events and people in it through the shared experience of suffering. In female-authored Romantic texts the connection and physical presence of the experiencing subject in the material world it describes also carries the additional function of lending authority to their voice, which can not, as in male-authored texts, be constructed with recourse to the figure of an autonomous poetic voice. While Byron's speaker, therefore, uses the construction of the "dream, which was not all a dream" vision to maintain poetic authority over the apocalyptic events, Browne's poem build's a sense of poetic authority on the situated physical experience of the speaker. Without the dream frame, this scenario would result in an absolute sense of inescapability for Browne's speaker, which means that the closing framework of the dreamer waking from her apocalyptic nightmare also functions as an exit strategy. As I have argued with O'Connell above, the Romantic understanding of dreams as prophetic as to the reality status of the events lends a sense of immediacy and inescapability to the nightmare scenario, as its effect carries over into the waking world of the speaker.

This residual materiality of the dream vision is underlined by the speaker's embodied connection to the physical world. While the explicitly constructed outlook position on the hill seems to create distance to the described world, this vantage point is necessary for the creation of a wide outlook anchored in concrete embodiment. In stanza 3, 4 and 11 the speaker's eyes move over the "far-spread heath", the "stately woods" that lie "farther on" and ultimately towards the "mighty Sea", cataloging the absence of movement. The spring, which "once had risen" from the heath had dried out as "the sultry glance of the sun, / And the glare of the dark blue sky, / Had checked its course, – no more to run."[93] Similarly, the woods in stanza 4, stand "like autumn wrecks [...] / Beneath the summer sky: / And every leaf, though dead, did keep / Its station in mockery; / For there was not one breath to sweep / The leaves from each perishing tree;" Like the dried-up spring, the trees are also "dead", their leaves "motionless [...] As though Death were too busy with other things / To sweep their corpses away."[94]

The personification of "Death" in line 33 works as a foreshadowing of more important (from an anthropocentric perspective) deaths in the following stanzas, which turn away from describing the dead, motionless natural world and towards the fate of "human creatures" before returning to the sea in stanza 11 and 12. This is an interesting parallel to "Darkness", which also ends its path of destruction on the "rivers, lakes and ocean [which] all stood still."[95] The stillness is again constructed

93 Browne, 156.

94 Browne, 156.

95 Byron, "Darkness," 717.

as an absence of water in Browne's poem as "the bed of the Sea" – also personified – "where he had slept" is visible, because all the water has dried up.[96] While in "Darkness" "nothing stirred within their silent depths",[97] the absence of water in "A World Without Water" does not only reveal the "treasures" of the sea but also "the monsters of the deep" which "lay dead, / With many a human form."[98] The description of the sea as an empty bed full of sailors' corpses, rare sea creatures and dead fish seems almost obscene, as "a myriad things, long lost and won, / Where there, unsought for, found". It is not death in itself that causes this impression, but the fact that the sea no longer serves as a "quiet bed / Away from the raging storm" and what should be hidden under the water's surface has now become visible, like the abject insides of a body unexpectedly exposed to the light of day.[99] The speaker's position towards this unusual spectacle emerges from the descriptions of disturbed rest, but perhaps most prominently in the use of "unsought for" in the stanza's last line. The poem envisions the personified, masculinized sea as the sleeping guardian of these unsought for treasures and horrors, the "last things" of a secularized apocalypse whose revelation does not include the possibility of redemption.

Where the sea is "sailorless" in "Darkness", Browne's poem, focused on human suffering and emotional connection, describes a last ship in stanza 12, "with all its sailors perishing / Even in the sight of land". The irony of the stanza is underlined by the lack of water, in the "very midst" of the ocean, "[w]here no human foot could pass." The stanza ends on the sailor's perspective, with the exclamation "Oh! Water had been a welcome sight / To that pale dying band!"[100] Unlike the water missing on land that is described in its functions as drinking water and to keep trees and plants alive, the sailor's desire for sea water is not entirely unambiguous, as the many dead bodies in the sea suggest. The exclamation "Oh!" at the beginning of the line underlines the paradoxical constructions of sailors deprived of the sea which aligns itself with the other forms of deprivation the poem imagines.

Community and Connections

While there is no sense of aesthetic sublimation in either "Darkness" or "A World Without Water", both poems frame their apocalyptic visions within the emotional range of humanity's reactions to the apocalypse. Agony is presented as a group sentiment through the use of "some", "the people" or even "the cor[p]ses", who "were

96 Browne, "A World Without Water," 158.

97 Byron, "Darkness," 717.

98 Browne, "A World Without Water," 158.

99 Browne, 158.

100 Browne, 158.

strewn / In the street, as dead leaves lay."[101] The use of this simile and the phrase "human creatures" introducing the turn to humanity's fate in line 36, suggest an interesting parallel between how the poem presents humans and non-human "nature" as similar in their suffering and death, as well as constructing a parallel between the formerly "stately wood" whose leaves "hung dead, motionless" and compared to the "cor[p]ses" in line 34 and the "forsaken" cities whose "marble wells were spent" and whose inhabitants' corpses did not feel "foul decay", but instead "withered alone" in the dry streets "as dead leaves" in line 50.[102] Incidentally, the poem dedicates its longest stanzas – stanza 4 of 12 lines and stanza 5 and 8 of 10 lines each to the description of dead trees and dying humans respectively, thus conferring a similar weight on the connected images ofhuman and non-human "cor[p]ses". The forsaken cities are not only the most prominent representative of (failed) human culture in the poem, their extraordinary status in a text which otherwise focuses predominantly on the effect of climate catastrophe on land- and seascapes is underlined by another blatant formal irregularity – the only line in the poem that does not end on a stressed syllable. The last, unstressed syllable of the word "forsaken" serves as an echo of lost (or rather abandoned) human culture, signaling that, while there is no escape from the drought in either natural or man-made environments, cities emerge as futile reminders of institutionalized culture as humanity searches for refuge in the natural environment.

As I have argued above, stanza 8 opens with an exclamation that dashes any expectations of relief or redemption: "Vain was the hope!" As the cloudless sky and "bright and beautiful [stars]" seem to almost mock the people's hopes of any sign of dew or precipitation. The rest of stanza 8 presents people's emotional reactions to their dashed hopes: "And women sat them down to weep / Over their hopeless pain: / And men had visions dark and deep, / Clouding the dizzy brain: / And children sobbed themselves to sleep, / And never woke again."[103] The wording of the men's reaction in particular suggests an intertextual connection to "Darkness", which can certainly be read as a "vision[...] dark and deep". The suggestion that this kind of vision might "[cloud] the dizzy brain" could, in this sense, be read as either an aesthetic or a content-related comment, but it does not emerge as particularly complimentary in either sense.[104] "Darkness", as I have argued above, focuses on isolation as one of the forms of apocalyptic deprivation it presents. The text denies any kind of connection between humans or humans and animals, treating even the bond between a dog and its master as ironic and useless. Instead, each person fights for themselves and dies alone. As I have argued above, Byron's poem does not distinguish between

101 Browne, 157.

102 Browne, 156–57.

103 Browne, 157.

104 Browne, 157.

men's and women's reactions, does not mention children at all and largely uses men's emotional responses to the apocalypse as examples for mankind. Browne's poem, instead, groups "people" and "human creatures" together, but also introduces an element of compassion in lines 35–6: "Oh terrible it was to think / Of human creatures then!"[105] The reactions described in stanza 8 must be read in light of this compassionate stance of the speaker towards them, who feels the women's "hopeless pain" and mourns the deaths of the children who, weaker than the women, but similar in their emotional reaction, "sobbed themselves to sleep, / And never woke again". This gendered focus on compassion is further supported by the suggested bond between mother and child in stanza 10:

A mother held her child to her breast
And kissed it tenderly,
And then she saw her infant smile;
What could that soft smile be?
A tear had sprung with a sudden start,
To her hot feverish eye;
It had fallen upon that faint child's lip
That was so parched and dry.[106]

The stanza's construction is overall sentimental and ironic, most prominently in showing that the mother's single tear will not be able to quench the child's thirst. The use of "tenderly" and "soft smile", however, still suggests a caring relationship between the mother and child that resists, even if only for an instance, the narrative of despair and hopelessness. Read in comparison with the "dark and deep" visions presented in "Darkness" and the solipsistic reactions to the apocalypse put forth in Byron's poem, Browne seems to suggest that a focus on women's responses would paint a different picture of care that is so deeply rooted in human connection, as to survive even beyond the "hopeless pain" of a waterless world without social structures (as the forsaken cities suggest) and beyond a hope of salvation. While it would be easy to dismiss this description of the mother and child as a sentimental overemphasis on traditional gender roles, the poem clearly juxtaposes this central image of care with the "desolate" and "lonely" position of "the very Sun" in the cloudless sky in the penultimate stanza: "And desolate seemed the very Sun, / He looked so lonely there!"[107]

The poem's use of male pronouns (common to most Romance languages), and capitalization for both the Sun and the Sea suggests not only a personification but

105 Browne, 156.
106 Browne, 157–58.
107 Browne, 159.

also an expectation of agency that surpasses human activity in the poem and positions light as a more destructive force than darkness, reversing the apocalyptic scenario presented in Byron's poem. While "Darkness" does not provide an explanation for the sudden extinguishing of the sun, "A World Without Water" does not only personify and capitalize the Sun, but it also underlines "his" agency and almost malicious intent, as, in stanza 9 "he" springs up "as if eager to see / What next his power could do" (creating a sense of more movement than any other entity throughout the poem exhibits).[108] In Browne's as in Byron's poem, the sun emerges as an important entity, but while "the bright sun" is extinguished early on in "Darkness" the personified, masculinized Sun becomes an agent of apocalyptic destruction in Browne's text. The speaker nevertheless seems to feel some compassion for the Sun, as "[h]e looked so lonely" in the sky. Browne's poem, accordingly, ends, not on the annihilation of life in light or darkness, but on compassion, as the speaker feels the "agony of thirst" that threatens to silence them: "I had a scorching swelling pain, / As if my heart would burst. / My tongue seemed parched; I tried to speak".[109] The irony of the voiceless speaker, who is only able to utter a "wild shriek" and, thus, wake from the dream, underscores the deep, compassionate involvement of the speaker, who is not a distanced observer as in Byron's "Darkness", but emotionally and physically immersed in the nightmare world. This marks perhaps the most important difference between "Darkness" and "A World Without Water", as the involvement of the lyrical persona in the events emerges much more prominently in Browne's text. The speaker feels thirsty along with everyone else and, in fact, is awakened from the nightmare by the individual sensation of thirst.

Beyond the speaker's physical reaction to the events, the two poems, thus, also differ considerably in tone and their creation of positionality, as I have argued above. Where "Darkness" creates a sense of distance between the humans experiencing the apocalypse and the speaker describing the events, Browne's speaker is physically present in the apocalyptic world and emotionally involved in its suffering, creating a sense of compassion. The use of interjections such as "oh" and exclamation marks emphasizing especially emotional observations throughout the text creates a sympathetic tone in "A World Without Water". The best example of the speaker's emotional involvement occurs in stanza 6, which frames the introduction of human experience in two exclamations, using "Oh" and "But alas" for emphasis:

Oh, terrible it was to think
Of human creatures then!
How they did seek in vain to drink
In every vale and glen;

108 Browne, 157.

109 Browne, 159.

And how the scorched foot did shrink
As it touched the slippery plain:
And some had gathered beneath the trees
In hope of finding shade;
But alas! there was not a single breeze
Astir in any glade![110]

The speaker's gesture of thinking "*of*" someone – of other humans – presents a prominent contrast to the construction of a universalizing subject position in canonized Romantic poetry, which Byron is able to reclaim via the dream vision in "Darkness", as I have argued above. The subject position of the speaker in Browne's poem is far less universal and there are several indicators of this throughout the poem from the spatiotemporal and embodied situatedness of the speaker to their emotional involvement. Instead of postulating the universality of the speaker's view, Browne's poem zooms in on various situated perceptions, creating compassion through the repeated use of exclamations (for instance, "Oh" and "Oh!" are repeated four times throughout the poem). The metonymical construction "the scorched foot" in line 39 is indicative of another strategy the poem uses to diversify the speaker's experience and draw the reader's compassion, as it uses the relatable physical sensation of walking on a hot surface, isolating it from the experiencing subject without universalizing the speaker's experience. Stanza 10 uses a similar strategy, narrowing the perspective to focus on the mother's "hot feverish eye" and the "faint child's lip" connected by the falling teardrop, but its most prominent use occurs in the transition from nightmare to waking in the poem's last stanza, which focuses on the speaker's own experience of thirst. In contrast to "Darkness", Browne's poem does not end in a proclamation of the universal, but rather a personal experience of the apocalyptic, as the speaker awakens parched and unable to fully voice this discomfort.

Given the fact that both "Darkness" and "A World Without Water" also describe the breakdown of a system, the disaster described in "A World Without Water" more closely resembles current scenarios of climate crisis and hints at human responsibility in the possible reactions to such climate events. Browne's focus on tracing the impact on human and non-human agents as well as evoking emotional responses to the devastation not only differs widely from the irony and emotional distance created in "Darkness", the poem's poetical construction also foregrounds humanity, community and compassion as values that might prevail even in the face of climate catastrophe. Mellor traces these differences back to Romantic models of understanding gender. In opposition to a binary construction of gender based on Otherness put

110 Browne, 156.

forth by male Romanticism, women writers of the Romantic period "typically endorsed a commitment to a construction of subjectivity based on alterity, and based their moral systems on [...] an ethic of care."[111] As Mellor suggests, "women writers of the Romantic period rejected this model of oppositional polarity (as the foundation of both the natural and the human worlds) replacing it with one based on sympathy and likeness" and "to fully understand their ways of knowing the world, we need to learn how to think *beyond* a dialectic based on polarities."[112] While Mellor herself acknowledges the "crude generalizations" and "limitations" of grouping and labeling the work of women writers of the Romantic period under the label of "feminine Romanticism",[113] any reassessment of such universal questions as the Romantics' relationship to the non-human natural world must acknowledge gendered differences in positionality and the construction of a poetic subject position, not so much as based in an essentialist binary model of gender, but rather in a gendered poetics emerging from social and cultural constructions of masculinity and femininity.

As I have argued above, both the deconstruction of the Burkean sublime in "Darkness" and the question of positionality in "A World Without Water" emerge from and respond to gendered cultural and aesthetic frameworks and gender constructions must, consequently, be taken into account as an underlying factor of how Romantic poetics envisioned the non-human natural world and potential subject positions from which to do so. While "A World Without Water" can be read as a response to "Darkness" in many ways that I have outlined above, the poem also presents a poetic expression in its own right – one that foregrounds relationality, community and care in ways that move beyond the solipsistic construction of the Romantic subject and, thus, enters into a dialogue with other texts by women writers of the Romantic period. These questions extend beyond genre boundaries, as I will show in the last part of this chapter, which focuses on Mary Shelley's novel *The Last Man*. As I will argue, the novel does not only engage with the poetic and aesthetic questions raised in "Darkness", it also foregrounds an ideology of chosen community and intellectual and emotional friendship in a multilayered framework of narrative strategies such as multi-perspectivity and -temporality that allow the narrator to engage with a variety of apocalyptic scenarios and model aesthetic reactions to them.

111 Mellor, *Romanticism and Gender*, 3.

112 Mellor, 4.

113 Mellor, 3.

Romantic Apocalypse in Mary Shelley's *The Last Man*

Compared to the two poetic texts I have discussed so far, *The Last Man* offers a less straightforward, if not less devastating, apocalyptic scenario – one that, paradoxically, leaves the non-human natural world largely intact while almost all of mankind is eradicated by a deadly pandemic.[114] This has far-reaching consequences for the ways in which the novel extrapolates situatedness and positionality on a number of narrative levels, as the narrator's isolated stance as the eponymous last man, contemplating his life retrospectively, mirrors Romantic constructions of poetic genius – the fine balance between "the spontaneous overflow of powerful feelings" and "emotion recollected in tranquility" proposed by William Wordsworth in the Preface to the second edition of *Lyrical Ballads* (1800) and reflected by Dorothy Wordsworth in the context of accessibility, as I have suggested in the "Steam" chapter above.[115] While the novel's main narrator, the eponymous "last man" Lionel Verney, does not establish himself as a poet or, indeed, a genius, his narrative explores the possibilities of expressing perceptions, experiences and emotions in the relationship between the writer and the natural world. More importantly, Verney's narrative is framed by and interspersed with contrasting experiences which draw attention to the impossibility of universalizing a single subject position even after literally eradicating all other experiencing subjects. While the novel engages with many of the aesthetic and representational questions that I have raised in the first two parts of this chapter, it does so within a different genre framework and in a substantially longer text that produces many of its ideas in a very gradual manner, thus, necessitating a substantially different analytical approach, that introduces the novel's complex narrative structure to frame its discussion of aesthetics and gendered positionality. My discussion of *The Last Man* will consequently address a number of interwoven concerns from genre and positionality, via the framing of narrative authority in the text to gender politics and the necessity of community and an addressee as the prerequisite of aesthetic experience and narrative. I will then move on to the novel's narrative framings of catastrophe and its aesthetics of disaster, before discussing its framing of community politics as a gendered disaster response and the ways in which the novel's narrative framework establishes a connection between prophecies and gendered positionality. The last part of my discussion concerns the novel's engagement with the failure of the sublime, which

114 There are striking parallels between the novel's treatment of the pandemic and official responses to the ongoing Covid-19 Pandemic in the Global North, which I will, however, refrain from highlighting, as they go far beyond the scope of my discussion here, even if there is a certain ironic dimension to reading, teaching and writing about pandemic literature in the midst of an ongoing pandemic.

115 Wordsworth, "Preface to Lyrical Ballads," 361.

is linked to the impossibility of a narrative position without community and an addressee. Throughout this discussion I will also address intertextual and thematic connections to "Darkness" and "A World Without Water", especially in terms of Shelley's engagement with Byron in the figure of Lord Raymond and in terms of the novel's questioning of sublime aesthetics and the construction of positionality and community as prerequisites to aesthetic experience and narrative production.

Genre and Positionality

Beyond its autobiographical and socio-political dimensions, as a response to personal loss,[116] *The Last Man* also functions as a reflection on the situatedness of writing, specifically regarding the manner in which the poetic subject relates to society and the non-human natural world. The text addresses the difficulties of making sense of and reacting to incomprehensible disasters – among them a pandemic, a solar eclipse, floods, meteor strikes and, ultimately, a depopulated earth, all of which reflect the complicated network of environmental consequences and coincidences witnessed by the Romantics in the wake of the eruption of Mount Tambora in 1815.

Verney, the last man, serves as a fractured mirror of Shelley's own concerns as a writer and editor of her husband's poetic legacy but also as an observer of natural and cultural phenomena that emerge as dramatic events in her novels. While *The Last Man* functions, in part, as a *roman à clef* that recasts Percy Shelley and Lord Byron in the characters of Adrian and Lord Raymond, the negotiation of political, social and literary ideas reaches beyond simple nostalgia into the realm of meta-commentary on Romantic ideals, aesthetics and literary communities.

As Paley argues, Shelley's *Last Man* presented itself as a relatively late entry in a literary tradition of "last-man"-narratives which had already descended into parody by the mid-1820s, a fact which contributed to the novel's lack of popularity and lukewarm critical reception.[117] As Yasmin Solomonescu has pointed out, for many

116 The plague narrative that emerges after a volume-long introduction of the protagonists and their personal and political struggles introduces the apocalypse as a tale of individual suffering and ultimate isolation that has often been read as an expression of Mary Shelley's own feelings of loss after the deaths of three of her children between 1815 and 1819, a miscarriage and the death of Percy Shelley in 1822, and the dissolution of her creative circle after her husband's death. Suparna Banerjee summarizes the autobiographical readings of the novel as a cathartic text, allowing Mary Shelley to recreate and reexperience "her maternal and other personal losses through a fantasy of apocalyptic destruction and disintegration that mirrors the chaos of 'her ruined inner world'", but concludes that "her act of mourning transcends the boundaries of the self. As a rich cultural text her grief-work comments on major socio-cultural themes that both informed the world she lived in and preoccupied her personally." (Banerjee, "Beyond Biography," 520.).

117 Paley, "Apocalypse Without Millennium," 12.

critics, "the chief problem" with the novel "was its failure to fix the reader's attention"[118] – an assessment which seems more than ironic for a text which, on a metanarrative level, is concerned with Romantic poetics and specifically the necessity of an addressee. This criticism was, however, not so much based on the novel's far-fetched events, but on its incoherence and its failure to create a consistent "aesthetic illusion",[119] suggesting that Shelley's contemporaries were more interested in being engrossed by a story than in the self-reflection of its production.

As the title page suggests, the novel was marketed as having been written "by the author of Frankenstein", thus attempting to draw on the fame and, to an extent also, notoriety of Shelley's earlier work and its popular theatrical adaptation *Presumption, or, the Fate of Frankenstein* by Richard Brinsley Peake, which had first been staged in 1823, further popularizing the novel. Unlike *Frankenstein*, which innovated the Gothic genre to the extent of determining its course for the nineteenth century and simultaneously laid the foundation for the Science Fiction genre, *The Last Man* drew on an already well-established tradition of last-man narratives, allowing its author to use the subject matter to also engage with Romantic poetics. While the earlier parodies had an impact on the reception of the novel, Shelley's work still takes the problem of "lastness" seriously on both an aesthetic and a personal level.[120] One of its central concerns is the connection between the Romantic writer, non-human nature reflected in Romantic writing and a real or imagined reader. The novel also complicates previous "last man" narratives in its adoption of a temporal framework that reaches both into the distant past and the far future by drawing on a prophetic, rather than a dream scenario in its frame narrative.

Verney's retrospective stance allows him to reflect his experiences of the end of his world, while his situation in time allows him to recreate a lost past in his narrative, but not a future, as his survival on an empty planet permits "no expectation of alteration for the better".[121] Instead, Verney's past spans several decades between the 2070s and the 2090s, functioning, to an extent, as a dystopian negotiation of present concerns projected into a far future for readers in 1826, by focusing on questions of government, social inequalities and restrictive gender roles, as well as broader ideas such as education, humanity, friendship, love and compassion. In spite of Shelley's exploration of the boundaries of scientific progress in *Frankenstein*, *The Last Man* is not a Science Fiction novel, as its futuristic England is described as a

118 Solomonescu, "Mary Shelley's Fascinations," 702.

119 Solomonescu, 703.

120 Another entry in the genre, a short story published anonymously under the initials "X.B." in *Blackwood's Magazine* in 1826, for instance, ends on a pun on "lastness", when the narrator is awakened from his nightmare of being the last man on earth by the entrance of his *man* servant, instantly dissolving a narrative of the end of the world into high farce (Anonymous, "The Last Man.").

121 Shelley, *The Last Man*, 469.

largely rural, agricultural society, in which people travel on horseback or in carriages and only occasionally use steamboats or air balloons to travel long distances. There are few signs of scientific advances or an industrial revolution – and certainly none of Blake's "dark Satanic Mills" – in the city of London, which is mainly described after its population has already been decimated by the pandemic.[122] This retro-futurism hinges, at least in part, on the narrator's perspective and experience of England as the rural idyll to the shepherd boy of his youth, but also on the novel's discussion of gradual improvements in the political system and the possibility of benevolent leadership. While, by the beginning of Verney's narrative, the monarchy has been abolished (peacefully by political process in the last generation), the country is still run by a landed gentry and their elected "Lord Protector" – a political position that plays an important role for the negotiation of ideals of political leadership oscillating between bitter monarchists, an inflexible republicanism and a form of benevolent stewardship that relies on personal dedication and community ethics more than it does on genuine democratic structures. The novel's main concern is, thus, not to construct a futuristic world or to juxtapose a rural and an industrialized society but to showcase Romantic ideals and their poetic expression[123] in a political climate that was less open to experiments than directly after the French revolution.[124]

This project explains the novel's foregrounding of emotions and experience, even beyond the possibility of expressing them. In what almost feels like a parody of the solipsistic Romantic poet, Verney, the last man on earth, vows to "witness all the variety of appearance that the elements can assume". But while he can "read fair augury in the rainbow – menace in the cloud – some lesson or record dear to [his] heart in everything",[125] these signs can carry meaning only so far as he decides to read it into them and remain meaningless in the absence of an addressee of his writing. Verney's gesture of ending his written account as he sets out on his lonely sea voyage "around the shores of deserted earth"[126] at the end of the novel is thus also a comment on the necessity of an audience that might serve as an addressee for his narrative in the future. In comparison to Byron's speaker, Verney emerges as overall less authoritative and always positioned in a clear situation of relationality to the events and his surroundings.

As the sole autodiegetic narrator of the novel's main diegetic level, Verney positions himself as an emerging writer early in his tale – an author dedicated to the

---

122 Blake, "Milton," 120.

123 Konstantinos Pozoukidis reads the novel's focus on the production of writing as a concern with "non-productive labor", that might, ultimately, be understood as part of an argument for the necessity of the humanities and their world-making abilities in the context of survival. (Pozoukidis, "The Survival of Non-Productive Labour in Mary Shelley's *The Last Man*," 393.)

124 see Hay, *Young Romantics*.

125 Shelley, *The Last Man*, 470.

126 Shelley, 470.

"biography of favourite historical characters, especially those whom I believed to have been traduced, or about whom clung obscurity and doubt."[127] While this might be read as a reference to Shelley's own laborious efforts of preparing her husband's work for publication and preserving his legacy after his untimely death in 1822,[128] Verney's contemplation of his authorial role also functions as a meta-poetic reflection on the role of the writer in a changing world:

> As my authorship increased, I acquired new sympathies and pleasures. I found another and a valuable link to enchain me to my fellow-creatures; my point of sight was extended, and the inclinations and capacities of all human beings became deeply interesting to me. Kings have been called the fathers of their people. Suddenly I became as it were the father of all mankind. Posterity became my heirs. My thoughts were gems to enrich the treasure house of man's intellectual possessions; each sentiment was a precious gift I bestowed on them. Let not these aspirations be attributed to vanity. They were not expressed in words, nor even reduced to form in my own mind; but they filled my soul, exalting my thoughts, raising a glow of enthusiasm, and led me out of the obscure path in which I before walked, into the bright noon-enlightened highway of mankind, making me, citizen of the world, a candidate for immortal honors, an eager aspirant to the praise and sympathy of my fellow men.[129]

Verney's position towards writing as a form of creation within a social framework that emerges from this passage explains his inability to continue his tale as the last man on earth. Without the sense of furthering human knowledge by "enrich[ing] the treasure house of man's intellectual possessions" for "posterity", his writing does not only become unmoored from its original purpose but is also presented as redundant in a world without readers. In its insistence on the role of the poet as a benevolent ruler (or "father of mankind"), the passage also shows traces of other discussions of Romantic authorship, most prominently, perhaps, Percy Shelley's declaration of the poet's creative power in "A Defence of Poetry" that "poets are the unacknowledged legislators of the world."[130] Mary Shelley's last man, however, at the same, time remains more humble in his assessment of this particular author's contribution, which he sees as "sufficiently unpretending",[131] and geared towards a view of the author as someone working for "[p]osterity" and the "praise and sympathy of my fellow men."[132] Unlike the more solipsistic poetic genius, incorporated for instance,

127 Shelley, 157.
128 see Hay, *Young Romantics*.
129 Shelley, *The Last Man*, 157.
130 Shelley, "A Defence of Poetry," 956. See also my discussion in the "Anthropocene" chapter above.
131 Shelley, *The Last Man*, 157.
132 Shelley, 157. I will address the gendered components of this distinction further below.

by both Wordsworth and Byron, Verney declares himself deeply interested in "the inclinations and capacities of all human beings", which extend his point of view and allow him to "enrich the treasure house of man's intellectual possessions."[133] It is interesting that Verney frames this discussion of his poetics as emotions "fill[ing] his soul", rather than a product of rational thought.

Framing Narrative Authority

The novel's retrospective narrative position lends an ironic tinge to any discussion of "posterity" in the absence of future readers and the narrator makes sure to acknowledge that this "account might as properly belong to a former period of [his] life as to the present moment."[134] As Andrew Sargent argues, Verney's position is indicative of the novel's "backward looking" view on the archive of Romanticism as something that exists in the past, but not in the present or the future.[135] This might also explain the novel's lack of a futuristic version of the future, as the last quarter of the twenty-first century remains very much a rural society – in spite of the rapid growth of industrialization in the first decades of the nineteenth century, as well as Shelley's own awareness of and engagement with new inventions and technologies in *Frankenstein*.[136]

A number of meta-narrative comments frame the novel's engagement with Romantic poetics, the emotions of the experiencing subject and the situatedness of narration. For Verney, his narrative functions as an "opiate, while it described my beloved friends, fresh with life and glowing with hope, active assistants on the scene, I was soothed"[137] and as a reflection of writing without an addressee: "I also will write a book, I cried – for whom to read? – to whom dedicated?"[138] He imagines the reader as a "tender offspring of the re-born world," a "fair being, with human heart, yet untamed by care, and human brow, yet unploughed by time"[139] who is admonished to,

> [l]et not day look on these lines, lest garish day waste, turn pale, and die. [...] seek some cave, deep embowered in earth's dark entrails, where no light will penetrate, save that which struggled, red and flickering through a single fissure, staining thy page with grimmest livery of death.[140]

---

133 Shelley, 157.

134 Shelley, 158.

135 Sargent, "Citation and the No Future of Romanticism in Mary Shelley's *The Last Man*," 313.

136 see Knellwolf, *Frankenstein's Science – Experimentation and Discovery in Romantic Culture, 1780–1830*.

137 Shelley, *The Last Man*, 267.

138 Shelley, 466.

139 Shelley, 437.

140 Shelley, 437.

This passage reflects Verney's own conflicting position towards his narrative, which he alternatingly uses as a refuge from the bleak, lonely present and as a catalyst of his own feelings. While he is unable to imagine continuing to write without at least an imagined addressee, the subject matter of his apocalyptic narrative is so unbearable for him on both a personal and a universal level, that he does not want it to see the light of day.

The admonition to bury the text underground echoes the novel's frame narrative, in which an unidentified autodiegetic narrator describes visiting Naples in 1818 (the year of the Shelleys' visit to Naples) and discovering the cave of the Cumaean Sybil, "a wide cavern with an arched dome-like roof. An aperture in the midst let in the light of heaven; but this was overgrown with brambles and underwood, which acted as a veil, obscuring the day, and giving a solemn religious hue to the apartment."[141] While the Shelleys were, according to their letters, unimpressed by the tourist attraction of the Sybil's cave, the unnamed frame narrator discovers "piles of leaves" and "fragments of bark" which were "traced with written characters. What appeared to us more astonishing, was that these writings were expressed in various languages: some unknown to my companion."[142] The frame narrator's claim of "deciphering these sacred remains"[143] and the thought that "obscure and chaotic as they are, they owe their present form to me, their decipherer" and, thus, "[d]oubtless the leaves of the Cumaean Sybil have suffered distortion and diminution of interest and excellence in my hands",[144] strikes a curious balance between a claim to authority over the narrative and the distancing found-manuscript trope common, for instance, in Gothic writing, but it also suggests a process of selection based on language and decipherability that lends an element of authenticity and unpredictability to the prophecy.

The frame narrative also functions as a meta-narrative commentary on Romantic poetics, as its deliberate construction of multiple, overlapping voices and selective editing of narrative fragments in various languages challenges the single-voiced narrative authority of the novel's main diegetic level. In the curious, semi-autobiographical frame narrative that highlights the voices of two women (the autobiographical author Mary Shelley and the mythical Cumaean Sybil), Verney's future voice is, ultimately, reduced to the "obscure and chaotic", "unintelligible" fragments of the Sybil's prophetic leaves and the deciphering work of the frame narrator.[145]

141 Shelley, 5.

142 Shelley, 5.

143 Shelley, 6.

144 Shelley, 6–7.

145 Melissa Bailes reads this construction as a "psychologization of geological catastrophe", arguing that the novel drew on geologist's attempts "to reconstruct fossilized remains of extinct species" to make sense of their extinction: "Just as her introduction temporally looks both backward and forward, Shelley explores various contemporary theories about what may have caused these past species' demise as a means of guiding her narrative anticipating human

Gender is, thus, one of the most prominent categories of narrative positionality that the novel addresses in its complicated frame structure and reflected throughout the text, as it painstakingly undermines the narrators authority and introduces his own sense of reflection, rather than invention.

While the frame narrative addresses the curious "subject of her verses, and the English dress of the Latin poet",[146] the tension between the autobiographical persona of a female, English expatriate writer in 1818 claiming to have found the autodiegetic narrative of an Englishman living in the late twenty-first century on some leaves left in a cave by an ancient Roman prophetess is never resolved in the novel. Timothy Ruppert reads this gesture as an indicator of Shelley's poetics of change, in which she surpasses her contemporaries: "What distinguishes *The Last Man* from like-minded Romantic calls for transformation is its use of the Sibyl, the revered pagan seer, as its principal visionary ancestress. By transferring vatic authority from the patriarchal Judeo-Christian scriptures to the matriarchal sibylline leaves, Shelley meaningfully questions the masculine bias in Romantic prophecy."[147] While I would agree with the general movement from male-connoted to female-connoted perspectivity in the frame narrative, I would also suggest that the texts resists a clear understanding of authority on several levels, undermining the idea of a clear transfer of authority to any of the narrative figures it introduces. Fuson Wang, by contrast, reads both the autodiegetic narrator of the frame narrative and the Sybil as deliberately "ungendered", but he also comes to the conclusion that "[g]ender and agency are Shelley's twin preoccupations in the introduction."[148] The construction of the frame narrative is, in any case, clearly deliberate, prefiguring the novel's complicated temporality and engagement with prophecy, as well as its exploration of gender roles in the split characters of the siblings Verney and Perdita, but also, to an extent, Adrian and Idris, as well as the gender-bending Evadne, whose role in the novel I will discuss below.

Gender Politics and the Necessity of Community

A large part of the novel is dedicated to the emotional reactions of its characters to the various circumstances and events that throw them together and, later, lead to the deaths of everyone but Verney. As in the two poems I have discussed above, individual fates within and beyond this group are presented as somewhat exemplary or

destruction." (Bailes, "The Psychologization of Geological Catastrophe in Mary Shelley's *The Last Man*," 671.) See also Kaitlin Mondello's reading of Shelley's novel as a "proto-evolutionary tale that anticipates Victorian and modern concerns about the definition and extinction of humans as a species" (Mondello, "The 'Grim Unreality,'" 163.).

146 Shelley, *The Last Man*, 6.

147 Ruppert, "Time and the Sibyl in Mary Shelley's *The Last Man*," 146.

148 Wang, "We Must Live Elsewhere," 237–39.

even models of humanity – albeit in more narrative detail. While agency – especially that of political leaders – is celebrated as highly commendable, the novel nevertheless also shows the futility of human endeavors in the face of events that lie entirely outside of human control, thus positioning itself in the field of (climate) dystopias, rather than utopias. The novel's sense of a fixed outcome, however, does not so much depend on fate but rather on the retrospective stance and many instances of foreshadowing of an outcome that is already fixed on the title page of the novel.[149] While the Sibylline prophecy introduced in the frame narrative seems to suggest a predetermined fate, the protagonists are still shown to struggle with their free will rather than a sense of acting out a fate predetermined by an agent of providence, suggesting that the novel principally serves, as Ruppert has argued, "to dramatize its author's theory of the human imagination as a legitimate source of both personal and public reformation."[150]

This program for change predominantly hinges on Verney's idealization of Adrian (modelled on Percy Shelley) as a true humanist, who is introduced as "wise and good".[151] His peaceful leadership emerges in several instances throughout the novel, but most prominently when he steps between two armies before they "shed [...] a drop of precious human blood".[152] While his contemporaries and political rivals (especially Raymond and Ryland) are focused on war and class struggle, he is the only person to see the Turkish casualties in the Greek war as "fellow being[s]" and "men and women".[153] Yet he is also shown to be impulsive, and senselessly lost at sea in a storm when the three remaining survivors of the plague – Verney, Adrian and Verney's niece Clara set sail from the Venetian Lido to Greece on a whim. The novel foregrounds this loss that seems, contrary to the many deaths the plague has caused at this point in the text, not only entirely preventable but particularly keenly felt by Verney who has idolized Adrian throughout the text (echoing Mary Shelley's own feelings of loss after Percy Shelley's death in a sailing accident). While the novel's character constellation prevents any kind of romantic entanglement beyond the homosocial between Verney and Adrian or any closer relationship between Adrian and Clara (who are modelled on Percy Shelley and their daughter Clara, who, in the text, is allowed to grow beyond her first year), the loss of this pair carries a gendered component in that they are idolized by Verney as models of both humanity and their gender. While Adrian is introduced in chapter 2 as a model of perfection,

---

149 On the function of temporality in this construction see Albright, "'In the Mean Time, What Did Perdita?': Rhythms and Reversals in Mary Shelley's *The Last Man*."

150 Ruppert, "Time and the Sibyl in Mary Shelley's *The Last Man*," 142.

151 Shelley, *The Last Man*, 29.

152 Shelley, 301.

153 Shelley, 161–62.

sensibility and courtesy,[154] Clara, the daughter of Lord Raymond and Verney's sister Perdita, who is raised by Verney and his wife Idris after her parents' deaths, grows into the epitome of romanticized womanhood caring for Verney and Idris' son Evelyn after his mother's death.[155] While there are several passages throughout the novel describing Clara's character and beauty, she is specifically confirmed in her gendered role as the "tenderest mother" in chapter 27, not long before her death at sea:

> Clara, [...] was now fourteen, and retained her childish appearance, though in height a woman; she acted the part of the tenderest mother to my little orphan boy; to see her playing with him, or attending silently and submissively on our wants, you thought only of her admirable docility and patience; but, in her soft eyes, and the veined curtains that veiled them, in the clearness of her marmoreal brow, and the tender expression of her lips, there was an intelligence and beauty that at once excited admiration and love.[156]

This passage does not only enumerate her beautiful features in an overly clichéd blazon, but also reiterates her patience, submissiveness and docility, which mark her as a model of womanly virtues.[157] The novel's narrative construction and Clara's unexplained melancholy complicate this verdict[158] – which is presented from Ver-

154 see Shelley, 26.

155 This endorsement of motherhood in Verney's narrative needs to be contrasted with the portrayal of gender in general, and mother figures in particular, throughout the text. Banerjee argues that in "Shelley's perception, then, woman – as lover and as mother, within the family or outside of it – is doomed to frustration, suffering and (self-)destruction. Shelley ascribes this to women's proclivity toward excessive dependence on emotional interaction with men and their inordinate absorption into motherhood." (Banerjee, "Beyond Biography," 526.)

156 Shelley, *The Last Man*, 417.

157 Wang argues that Clara is also presented as an idealized figure of whiteness, as "paler than the whitest snow" (Shelley, 337.), in one of the few racialized episodes of the novel (see Wang, "We Must Live Elsewhere," 242.), which contrasts Verney's encounter with an infected, "half clad" Black man who "writhing under the agony of disease" and held him "with a convulsive grasp" while "his breath, death-laden, entered my vitals" (Shelley, *The Last Man*, 336–37.) with his discovery of his older son Alfred's death and the beginning disintegration of his family. The encounter with the Black man leads to Verney's infection but it also underlines the novel's narrative of "racialized immunity" as the breath of the Black man serves to inoculate Verney against the plague (Wang, "We Must Live Elsewhere," 242.). The unexplained sudden appearance and disappearance of the nameless Black man underlines the novel's strategy of highlighting the experiences of its main characters and select side characters as exemplary, while instrumentalizing the fate of the dying (in many cases racialized) multitudes to heighten the emotional struggles of the main characters.

158 Reading Clara in the context of the novel's presentation of gender roles in the contemporary social context, Banerjee draws attention to her melancholy as an expression of her rejection of a role as a future mother of Adrian's children, suggesting that "her strong accents in her

ney's perspective – as gender roles are split across several characters throughout the text. This specific passage functions as a reiteration of Clara's virtues specifically in conjunction with her role as de facto mother to Verney's son Evelyn just before the child's death and before she and Adrian – whose virtues Verney has equally praised throughout the novel – are lost at sea in chapter 29.

This fatal chapter, which turns Verney into the last man, opens on a lengthy contemplation of the writer's position and the task of finishing a narrative without hope or an addressee. The retrospective narrative perspective is steeped – quite openly – in "black Melancholy" and "solitude" which overpower non-human nature in the narrator's consciousness:

> Arise, black Melancholy! [...] Bring with thee murky fogs from hell, which may drink up the day; bring blight and pestiferous exhalations, which, entering the hollow caverns and breathing places of the earth may fill her stony veins with corruption, so that not only herbage may no longer flourish, the trees may rot, and the rivers run with gall – but the everlasting mountains be decomposed, and the mighty deep putrify, and the genial atmosphere which clips the globe, lose all powers of generation and sustenance. Do this, sad visaged power, while I write, while eyes read these pages.[159]

There are distinct echoes of Shakespeare – predominantly of *Macbeth* and of a corrupted version of the meta-poetic closing couplet of sonnet 18 – in this passage, deepening an intertextual connection that the novel established earlier, when Verney saw one of the last performances of the Scottish play on a visit to London in chapter 19, and drawing attention to the novel's similar construction of a prophesied downfall. As Macbeth, Verney is also to an extent fulfilling a cryptic prophecy that straddles the boundary of fate and free will. His invocation does not only address a power that, as he immerses himself in it, is able to destroy the whole of the planet, but it also links this destruction to the task and process of reading and writing in the meta-narrative construction of "while I write, while eyes read these pages". Verney's words are so toxic to his own mind that he must warn future generations of readers to beware of their destructive power, but, in his role as the last man, he also remains entirely powerless, as there are no "eyes [to] read these pages" but his own. The absence of a reader, thus, negates the performative power of the invocation, which, like Verney's other outbursts remains powerless. Moreover, Verney's perspective returns to the entanglement not just of human perspective but of literary perspective with the existence of the/his world over and over, frequently asking whether the planet in all its glory can continue to exist without an observer:

---

final statement seem almost to prefer the eternal but asexual embrace of Adrian in death than a life as his wife." (Banerjee, "Beyond Biography," 528.).

159 Shelley, *The Last Man*, 437.

> Will the earth still keep her place among the planets; will she still journey with unmarked regularity round the sun; will the seasons change, the trees adorn themselves with leaves, and flowers shed their fragrance, in solitude? Will the mountains remain unmoved, and streams still keep a downward course towards the vast abyss; will the tides rise and fall, and the winds fan universal nature; will beasts pasture, birds fly, and fishes swim, when man, the lord, possessor, perceiver, and recorder of all these things, has passed away, as though he had never been?[160]

I have discussed an extract that raises similar questions in the "Anthropocene" chapter above, but this passage raises farther reaching questions about the role of humanity in the context of the non-human natural world, as it frames "man" as "the lord, possessor, perceiver, and recorder of all these things". The passage seems to echo Percy Shelley's closing question in "Mont Blanc" ("And what were thou, and earth, and stars, and sea, / If to the human mind's imaginings / Silence and solitude were vacancy?"[161]), although shifting the focus from human indifference to human absence. For both Shelleys the non-human natural world is constructed in the perspective of the poet, but for Mary Shelley's Verney this construction also requires the connection to fellow minds with which to share this aesthetic experience. Her narrator figure, thus, emerges less as a reconstruction of the Romantic poet as lone genius, but rather as a reflector of political, cultural and aesthetic theories presented by the other characters, most prominently Adrian (modeled on Percy Shelley).

It is interesting that, beyond the question of human dominance over "nature" ("the lord, possessor") in this passage the observation of "nature" is once again connected to the act of writing, as "man" is not only cast as the "perceiver" but also as the "recorder of all these things". As Olivia Murphy observes, the text seems to suggest that "the answer to these questions is, implicitly, *yes*",[162] however, in the light of Verney's observations about the power of narrative as destructive force, this is also a comment on "nature" as a construct of human perception and poetic production. Murphy reads *The Last Man* from a twenty-first-century, anthropocene perspective as a comment on the ultimate expendability of mankind for a "world that will not miss its human denizens" in all their unsustainable, destructive force.[163] *The Last Man* is, as Richardson argues, "uniquely interested in representing the history of human civilization and the relation between human populations and the environment",[164] paving a path for a number of ecocritical readings in the last few years.

---

160 Shelley, 412–13.

161 Shelley, "Mont Blanc," 849.

162 Murphy, "Apocalypse Not Quite," 253.

163 Murphy, 253.

164 Richardson, "The Environmental Uncanny: Imagining the Anthropocene in Mary Shelley's *The Last Man*," 1062.

Framing Catastrophe

Verney's comments do, however, not just center on the potential solitude of "nature" without humanity but on the necessity of a human observer to produce the specific poetic *image* of "nature" that Verney has constructed throughout the text and that will perish with the figure of its "recorder". As he observes in comparing himself to Robinson Crusoe in chapter 29, "I alone bore human features; I alone could give articulation to thought; and when I slept both day and night were unbeheld of any."[165] What may sound like a god complex, is, ultimately, an expression of poetics as Verney's position as the last man, writer and poetic subject on the planet renders both "nature" and writing dependent on his powers of observation. Unlike the solipsistic figure of the Romantic poet as genius, Verney does not derive aesthetic pleasure from this experience, but instead uses it to lament the absence of likeminded companions with whom he might share the experience, thus negating the idea of sublimation through the individual experience of the autonomous self.

While Verney is the only narrator on the novel's main diegetic level, his narrative perspective still includes events experienced by other characters in his absence that are later reported to him, and he frequently juxtaposes framings of events within the scope of different contemporary discourses. These include questions of leadership and pandemic responses (the explicit differences between Adrian's, Raymond's and the opposition leader Ryland's), but also individual or group-specific reactions to natural disasters. In these instances, the novel negotiates possible reactions to climate catastrophe alongside explorations of the perceiving and writing subject's engagements with the non-human natural world which are steeped in but also, on occasion, challenge Romantic poetics. Verney's role as an observer is, in these instances, supplemented by the reported reactions of other perceiving subjects, whose varying sensibilities embody different responses to a plethora of seemingly connected, but ultimately unexplained events.

The novel's apocalyptic narrative is based on a web of intertwined smaller catastrophes, whose main function in the text seems to be to raise questions of perception and appraisal, while most attempts at making sense of these events are presented as overly narrow or biased, thus raising questions about the possibilities of perceiving and making sense of (climate) catastrophe. The most striking example of how the novel presents these questions is an extended discussion of the pandemic response tinged with hubris, British nationalism and the presumption of Western/European superiority alongside religious framings of various unusual cosmic events.[166] Ver-

165 Shelley, *The Last Man*, 448.

166 As Ranita Chatterjee has argued, the politics of infection and the responses to the pandemic described in the novel can be read in the context of Agamben's biopolitics (Chatterjee, "Our Bodies, Our Catastrophes," 35.).

ney offers several discussions of the encroaching climate catastrophe throughout the novel, with the clearest description occurring at the beginning of chapter 16, where he quotes Hesiod to supplement his own assessment with what seems, at first glance, a religious reading:

> Some disorder had surely crept into the course of the elements, destroying their benignant influence. The wind, prince of air, raged through his kingdom, lashing the sea into fury, and subduing the rebel earth into some sort of obedience.
>
> The God sends down his angry plagues from high,
> Famine and pestilence in heaps they die.
> Again in vengeance of his wrath he falls
> On their great hosts, and breaks their tottering walls;
> Arrests their navies on the ocean's plain,
> And whelms their strength with mountains of the main. [Hesiod]
>
> Their deadly power shook the flourishing countries of the south, and during winter, even, we, in our northern retreat, began to quake under their ill effects.[167]

There is a curious contrast of secular and religious framings in this passage, which emerges from the juxtaposition of Verney's words and the chosen quotation. Verney's understanding of the world as made up of "elements" with a generally "benignant influence", which he enumerates in the following sentence – air, water and earth making an appearance in the form of winds, sea and, interestingly, "rebel earth" (suggesting a reference to human agency and willfulness) – clashes with Hesiod's narrative of an angry god, tormenting humanity in the name of vengeance. The pre-Christian framing lends an interesting component of materiality to this act of revenge, as Hesiod's Zeus (and other Greek deities) are known to interfere in human lives to a larger extent than the creator-Gods of the later monotheistic religions. The quotation is followed by a framing of current events in the context of global political entities. While here "north" and "south" refer to England (introduced as an isolated island retreat on the first page of the novel) and the "flourishing" countries of southern Europe (predominantly France, Italy and Greece), rather than the more current meaning of Global North and Global South, the novel includes England's view of non-European countries elsewhere, most prominently in the discussion of the country's surmised immunity to the plague in chapter 16. In this passage, Verney's solitary perspective broadens to include the general opinion of his countrymen:

> Can it be true, each asked the other with wonder and dismay, that whole countries are laid waste, whole nations annihilated, by these disorders in nature? The vast cities of America, the fertile plains of Hindostan, the crowded abodes of the

167 Shelley, *The Last Man*, 229.

> Chinese, are menaced with utter ruin [...] As yet western Europe was uninfected; would it always be so? In the face of all this we call ourselves lords of the creation, wielders of the elements, masters of life and death, and we allege in excuse of this arrogance, that though the individual is destroyed, man continues for ever.[168]

The scale on which Verney discusses both the possibility of annihilation and of survival throughout the novel is the nation, suggesting that when a whole nation is threatened "man [...] feels his tenure of life insecure, his inheritance on earth cut off."[169] These passages, which present a position beyond Verney's perspective, extending to a "community" rattled by the news of disasters, also reflect a colonialist/white supremacist sense of superiority of the English, who are generally believed to be immune to a "[p]lague" that "drinks the dark blood of the inhabitant of the south, but [...] never feasts on the pale-faced Celt."[170] The phrasing is suggestive of a society that does not want to face the real danger and instead relies on mutual reassurances, but also draws attention to the deep-seated jingoism and colonialist mentality of English society which had "[l]ate [...] envied their abodes, their spicy groves, fertile plains, and abundant loveliness",[171] but now expects to remain uninfected by a virus decimating every other population on earth, lamenting their deaths but mainly concerned with the effects of this calamity on the "prosperity of the nation".[172] Verney's ambiguous perspective in these passages does not entirely adopt or reject these positions, which complicates his criticism.[173] His narrative position adds a sense of fatality and distance to these observations, presenting a retrospective view of political discussions and responses to the pandemic from his position as an atemporal observer. While he groups himself among the English, he does not possess the political ambition of Raymond, Ryland or Adrian and merely supports the others' efforts to fight for "freedom" or sustain the social order out of friendship and admiration.

To Verney, returning from the war in Greece, where he has lost Raymond and his sister, in 2092, his home at Windsor seems "the abode of every kindly spirit. Hap-

---

168 Shelley, 233.

169 Shelley, 230.

170 Shelley, 233.

171 Shelley, 233.

172 Shelley, 234.

173 Emily Lyons reads the opening of Verney's narrative as an expression of his worldview, which is based on "the preeminence of English culture" as "actually inscribed on the land itself, which is apparent in the direct connection he draws between the significance of the physical landscape of England and the 'mental power' of its natives." (Lyons, "O Little Isle!," 473.). However, Ruppert argues against "reading her novel primarily as an indictment of British imperialism" as "many Romanticists fail to note that Shelley presents a unique vision of solidarity in the face of global catastrophe and so neglect the narrative innovations through which she crafts that vision." (Ruppert, "Time and the Sibyl in Mary Shelley's *The Last Man*," 142.)

piness, love and peace, walked the forest paths, and tempered the atmosphere" and after, "hav[ing] lived" he decides to "shut the door on the world."[174] While, for Verney, this sentiment is related to both his surroundings and his family, it is also framed as political by Adrian's assessment of "peace throughout the rest of the world", as most countries are trying to contain the pandemic:

> 'Let this last but twelve months,' said Adrian; 'and earth will become a Paradise. The energies of man were before directed to the destruction of his species: they now aim at its liberation and preservation. [...] What may not the forces, never before united, of liberty and peace achieve in this dwelling of man?'[175]

Throughout the text, Adrian is presented as a benevolent, humanist leader, whose main aim is the preservation of mankind. While Ryland, his main opponent, frames him as a dreamer, he also suggests that struggle and strife are natural occurrences and rooted in human nature: "When the seasons have become equal, when the air breeds no disorders, when its surface is no longer liable to blights and droughts, then sickness will cease; when men's passions are dead, poverty will depart."[176] While Verney tends towards Adrian's political views and Ryland is later shown to be a coward in the face of the raging pandemic, Verney's narrative juxtaposes these two opinions without resolution, instead opting to have the discussion interrupted by news from Greece about the spread of the pandemic. Verney's function in these chapters is the presentation of emotional reactions to the approaching apocalypse. While he frequently interrupts his narrative with his own emotional framings of the events, here he withdraws into the background, creating a distance between himself as the only survivor and the (failed) survival strategies emerging along the way. As an observer, rather than a leader, he can report on these incidents in a distant manner, but his narrative distance also extends to the non-human natural world and a number of apocalyptic events that coincide with the pandemic.

The Aesthetics of Disaster

A first framing of "nature" as no longer benevolent occurs in chapter 16, where the narrative describes her as an all-powerful mother, about to turn on her children:

> Nature, our mother, and our friend, had turned on us a brow of menace. She shewed us plainly, that, though she permitted us to assign her laws and subdue her apparent powers, yet, if she put forth but a finger, we must quake. She

174 Shelley, *The Last Man*, 218.

175 Shelley, 219.

176 Shelley, 219–20.

could take our globe, fringed with mountains, girded by the atmosphere, containing the condition of our being, and all that man's mind could invent or his force achieve; she could take the ball in her hand, and cast it into space, where life would be drunk up, and man and all his efforts for ever annihilated.[177]

The end of this passage bears a striking similarity to "Darkness", and *The Last Man* also repeats the poem's gendering of the destructive force as female, while identifying it with the figure of "nature" as a usually benevolent mother, who has now "turned on us a brow of menace". The personification of "nature" as "mother" and "friend" sends mixed signals of support and menace, that are here juxtaposed to the utter indifference of the universe in Byron's "Darkness". This ambiguity highlights "nature"'s agency, but also positions her as somewhat volatile, posing a threat to humanity. The realization of being at the mercy of this all-powerful figure diminishes human "force" and inventiveness, but, curiously, does not strengthen an understanding of humanity's position as a *part* of this "nature", instead introducing a third, common entity – the planet – and separating "mother nature" not only from humanity, but also from "our globe". Unlike in the previous chapter, this sentiment is no longer ascribed to a single character but presented by a universalizing "we": "[t]hese speculations were rife among us" and "[w]e could not fear – we did not. Yet a feeling of awe, a breathless sentiment of wonder, a painful sense of the degradation of humanity, was introduced into every heart."[178] The aesthetical framing of these thoughts as inspiring "awe" and "wonder" at the destructiveness of "nature" suggests another parallel to Byron's poem, which, as I have argued above, evokes the sublime as a possible category of aestheticizing climate catastrophe but ultimately also discards it in the face of the "degradation" of humanity in the apocalypse which offers no possibility of aesthetic sublimation. *The Last Man* takes up a similar stance here in evoking "awe" and "wonder" without the distance necessary to make the "degradation of mankind" a sublime experience. While, in this view, human inventiveness and the ability to scientifically describe "nature" envision humanity as out of reach of disaster, this passage serves as a reminder (directed by Verney at himself and his fellow human beings) of the fragility of this stance in the universe. Verney's position as a retrospective observer does not allow for any sense of sublimation, as he is not only fully aware of the outcome of all the events he narrates, but also presents himself as an observing, rather than an experiencing/acting subject.

The novel evokes an even stronger, yet equally distant, echo of "Darkness" in the presentation of an eclipse in chapter 15, interestingly framing reports of the event as incredible – if not for their numbers – and human reactions to the event as excessive in their irrational fear and religious fervor:

177 Shelley, 232.

178 Shelley, 232.

> A strange story was brought to us from the East, to which little credit would have been given, had not the fact been attested by a multitude of witnesses, in various parts of the world. On the twenty-first of June, it was said that an hour before noon, a black sun arose: an orb, the size of that luminary, but dark, defined, whose beams were shadows, ascended from the west; in about an hour it had reached the meridian, and eclipsed the bright parent of day. Night fell upon every country, night, sudden, rayless, entire. The stars came out, shedding their ineffectual glimmerings on the light-widowed earth. But soon the dim orb passed from over the sun, and lingered down the eastern heaven. As it descended, its dusky rays crossed the brilliant ones of the sun, and deadened or distorted them. The shadows of things assumed strange and ghastly shapes. The wild animals in the woods took fright at the unknown shapes figured on the ground. They fled they knew not whither; and the citizens were filled with greater dread, at the convulsion which "shook lions into civil streets;" – birds, strong-winged eagles, suddenly blinded, fell in the market-places, while owls and bats shewed themselves welcoming the early night. Gradually the object of fear sank beneath the horizon, and to the last shot up shadowy beams into the otherwise radiant air. Such was the tale sent us from Asia, from the eastern extremity of Europe, and from Africa as far west as the Golden Coast.[179]

The description of an eclipse is framed as incredulous from the start, but the fact that there were many witnesses to it is repeated at its beginning and end, with the geographical details adding both credibility and contributing to the English sense of superiority at hearing these outlandish superstitions. While Verney's narrative seems focused on reporting the strange occurrences in all possible accuracy – including the use of quotations – there is a certain sensationalism throughout the passage which resonates with the idea of "strangeness". While the passage studiously avoids the word "darkness" – using instead words associated with it, such as "black", "dark", "night" and "shadow" – there are, in fact, a number of echoes of Byron's "Darkness" – most prominently the use of "rayless", "blinded", and the description of wild and dangerous animals fleeing the woods and invading the "civil streets", not to mention the event itself. The date of the eclipse, 21 June, establishes a further connection to Byron, as "Darkness" was included in *The Prisoner of Chillon*, a text which Byron had begun to compose on a sailing expedition with Percy Shelley to the Chateau de Chillon on 22 June 1816, while Mary Shelley was working on *Frankenstein* between producing fair copies of Canto III of *Childe Harold's Pilgrimage* (1812–18) for Byron.[180]

The passage in its entirety oscillates between incredulity and a rational scientific explanation, seeming to suggest a tongue-in-cheek comment on the overreaction to the combination of sunspots and gloomy weather that Byron's poem might be read

179 Shelley, 223–24.

180 see Stabler, "1816 in the Manuscripts of Lord Byron and Mary Shelley."

as from Mary Shelley's extremely contemporary perspective. The description of the different animals' reactions to the eclipse described in the passage is particularly interesting, as it seems to suggest a plurality of perspectives that is absent from Byron's text. As if to reintroduce the concept of nocturnal animals, "owls and bats shewed themselves welcoming the early night", drawing attention to the possibility of an alternative interpretation of the events that moves beyond the common panic reactions of humans and animals dependent on daylight.

This critical note is further supported by Verney's description of the far-reaching climate of panic following the relatively brief event: "Through Asia, from the banks of the Nile to the shores of the Caspian, From the Hellespont even to the sea of Omar, a sudden panic was driven. [...] The plague was forgotten, in this new fear which the black sun had spread."[181] As the "tale of the sun of darkness" – another reference to Byron's text – spreads further West, reactions are driven by fear and renewed religious fervor, regardless of belief system. Consistent with its general atheist approach, the text groups "christians" [sic] among other religions, thus suggesting that all religious responses are similarly futile and, to an extent, dangerous in rushing the spread of the pandemic.[182] While the solar eclipse is, in itself, framed as unrelated to the pandemic, the ensuing panic has an impact on the number of deaths and the global spread of the plague, which arrives in England with English refugees from the East,[183] thus contradicting the earlier framing of the English as immune to a pandemic decimating the populations of other countries. As "Darkness", *The Last Man* also avoids a numinous framing of the events, insisting instead on narrating the ways in which different human actions combined in creating specific effects.

The novel presents a number of other natural disasters as similarly entwined with the plague, as one heightens the impact of the other. In Verney's narrative the plague takes several years to spread through northern Europe and England, decimating the population and diminishing the work force to an extent that infrastructure is neglected, but the effects are, again, first visible elsewhere:

> In Italy, the rivers were unwatched by the diminished peasantry; and like wild beasts from their lair when the hunters and dogs are afar, did Tiber, Arno, and Po, rush upon and destroy the fertility of the plains. Whole villages were carried away. Rome, and Florence, and Pisa were overflowed, and their marble palaces,

181 Shelley, *The Last Man*, 224.

182 Shelley, 224.

183 see Shelley, 225. There is a certain irony in this construction of the English infecting the English which, from a current point of view, countermands early xenophobic framings of the ongoing Covid-19 pandemic as a specifically East-Asian disease (with the predictable consequence of Anti-Asian racist violence).

> late mirrored in tranquil streams, had their foundations shaken by their winter-gifted power. In Germany and Russia the injury was still more momentous.[184]

In the rural societies the novel describes, fields remain untended when the population is diminished, and panic reactions drive city-dwellers to the country and the country population to the city.[185]

Contrasting Communal and Solipsistic Responses

While the novel showcases a variety of coping strategies, none of them are ultimately successful in doing more than diminishing immediate harm to individuals and upholding the humanity of the main characters. One of the most interesting episodes in this context is the interspersed narrative of Martha, an old woman whose survival of an earlier plague positions her as a community leader of the village of Little Marlow:

> The plague entered the village; and while fright and grief deprived the inhabitants of the little wisdom they possessed, old Martha stepped forward and said – 'Before now I have been in a town where there was the plague.' – 'And you escaped?' – 'No, but I recovered.' – After this Martha was seated more firmly than ever on that regal seat, elevated by reverence and love. She entered the cottages of the sick; she relieved their wants with her own hand; she betrayed no fear, and inspired all who saw her with some portion of her own native courage.[186]

While, unlike some of the other embedded narratives, Martha's story is not revisited, her example of a politics of care is, nevertheless, described as inspirational, suggesting to Verney a plan to "go from village to village, seeking out the rustic archon of the place, and by systematizing their exertions, and enlightening their views, increase both their power and their use among their fellow-cottagers."[187] The gendered framing of this episode is interesting, as it presents Martha not only as a matriarch whose wisdom is based on her care for others, but also as someone who is revered by her community as "regal". Beyond her function of introducing a model for a community-based politics of care, Martha's example also echoes the novel's focus on the complexity of subject positions which is negotiated via a number of interspersed narra-

184 Shelley, 269.

185 As Wood argues, early nineteenth-century economies relied largely on subsistence farming, which the cold and wet summers after the eruption of Mount Tambora disrupted with devastating effects. The Shelleys became aware of this problem after their return to England in the summer of 1816 (see Wood, *Tambora*.).

186 Shelley, *The Last Man*, 271.

187 Shelley, 272.

tives that specifically question authoritative and oversimplified models of explaining the world. Martha leads by example, not through any specific power, suggesting an equalizing, if not democratizing effect of common suffering. This is especially relevant, as Verney uses the opportunity to juxtapose her description with that of the inhabitants of "manor houses and gentlemen's seats" he has visited and who he found "actuated by the purest benevolence and ready to lend their utmost aid for the welfare of their tenants", but also dismissing their efforts as "not enough".[188] This curious combination of praise and criticism aligns itself with Adrian's self-effacing model of leadership and Verney's attempt to support his friend's role as the Lord Protector of a dying population in all things possible.

A sense of progression links the catastrophic events, which also seem to move closer to the protagonists (starting in the East and moving across Europe to England) alongside the pandemic. The early assessment of the uselessness of religious myths of apocalypse and redemption is repeated throughout the text, most prominently when another strange cosmic phenomenon, the appearance of meteors during a flood, is greeted with fear, as well as religious explanations. As with the solar eclipse, Verney's narrative of the event is very detailed, creating an initial sense of distance between the chronicler of the events and their very clear danger that is, however, in this case abandoned for a more direct description of his own emotional reaction:

> All day we watched the raging of the mountainous waves, and towards sunset a desire to decypher the promise for the morrow at this setting, made us all gather with one accord on the edge of the cliff. When the mighty luminary approached within a few degrees of the tempest-tossed horizon, suddenly, a wonder! three other suns, alike burning and brilliant, rushed from various quarters of the heavens towards the great orb; they whirled round it. The glare of light was intense to our dazzled eyes; the sun itself seemed to join in the dance, while the sea burned like a furnace, like all Vesuvius a-light, with flowing lava beneath.[189]

The description does not give any explanation for this phenomenon[190] but initially frames it with a sense of "wonder" (with an exclamation mark) and in the imagery of volcanic eruption. The timing of the event at sunset, during a raging storm which has flooded the streets of Dover adds an additional gravity and meaning to its appearance, as does the fact that the witnesses are the last survivors of the pandemic on British soil who are planning to leave for the continent in search of a warmer

---

188 Shelley, 272.

189 Shelley, 370.

190 As I have argued above, this is typical of literary framings of climate events in the wake of the eruption of Mount Tambora, as the Romantics were unaware of what triggered the strange, sometimes even unconnected climate phenomena they witnessed.

climate in the south. While their exodus is triggered by the pandemic, the group are essentially climate migrants searching, as Verney puts it, for "some natural Paradise, some garden of the earth where our simple wants may be easily supplied, and the enjoyment of a delicious climate compensate for the social pleasures we have lost."[191] This is, of course, an anachronistic reading, but the novel's earlier evocation of xenophobia and British superiority makes this colonial move sound particularly ironic.

The novel's framing of the white cliffs of Dover as a historical end point supports the sense of liminality surrounding them, highlighting the juxtaposition of a number of notable binaries (night and day, fire and water, land and sea, as well as home and abroad), which further the impression of a life/death scenario. The narrative presents the different reactions to the events, from Verney's own assessment that the additional "mock suns" are in fact "meteors", to terror and panic among the horses and cattle who "raced down the brink of the cliff, and blinded by light, plunged down with frightful yells in the waves below."[192] The spectacle is immediately followed by a tsunami-like wave rising towards the setting sun and then moving towards the shore: "[I]t seemed that the sea rose to meet it – it mounted higher and higher, till the fiery globe was obscured, and the wall of water still ascended the horizon." The combination of meteors and the approaching "wall of water" leads to further panic reactions among the watchers, which Verney again describes in detail:

> Many cried aloud, that these were no meteors, but globes of burning matter, which had set fire to the earth and caused the vast cauldron at our feet to bubble up with its measureless waves; the day of judgment was come, they averred, and a few moments would transport us before the awful countenance of the omnipresent judge;[193]

The subject position changes throughout this passage beginning with a vague "many", towards a "they", only to include the narrator again in the "us" in the last sentence, reflecting Verney's ambiguous involvement as an observer of the event, who does, however, not share the religious premonitions. The verb "averred" underlines Verney's disbelief as well as the tenacity of "them". The assessment is immediately framed as having been given by "those [...] given to visionary terrors",[194] while Verney's own response is framed within the canon of Romantic aesthetics with a heavy dose of fatalism: "A sublime sense of awe calmed the swift pulsations

191 Shelley, *The Last Man*, 312.

192 Shelley, 370.

193 Shelley, 371.

194 Shelley, 371.

of my heart – I awaited the approach of the destruction menaced, with that solemn resignation which an unavoidable necessity instils."[195]

This expectation of the end aligns itself with Verney's earlier observations of the black sun. The wall of water thus also functions as a reminder of an imminent global apocalypse that is framed in the imagery of "Darkness": "[I]t appeared as if suddenly the motion of earth was revealed to us – as if no longer we were ruled by ancient laws, but were turned adrift in an unknown region of space."[196] The novel thus does not only juxtapose the religious readings of the crowd with Verney's fatalism, but it also draws on the sublime as a failed aesthetic category in the face of apocalypse. Verney is not awed by a cataclysmic spectacle viewed from a safe distance, but by what he sees as an "unavoidable necessity" of the end. As in "Darkness", Verney's narrative in *The Last Man* suggests that there is no position of safety in the apocalypse, even though, as the only survivor, he has the safest position in it. This paradoxical situatedness is a consequence of the novel's complicated temporal structure, in which, as Richard Albright argues, a "summary/scene rhythm" envelops "a series of stories within Verney's account, which interrupt the flow of the narrative and draw us backward."[197] I have discussed one example – Martha's story – above; the second concerns Juliet, a young woman who, after a series of calamities in England, is drawn into the circle of a false prophet in Paris only to die of the plague alongside him and his "congregation of the elect", thus "destroy[ing] the illusion" and proving his inadequacy as a prophet.[198] The third example revolves around Lucy Martin, who is introduced as another model of care in response to the pandemic, when she refuses to leave her ailing, elderly mother behind as everybody else is preparing to leave the country. Lucy Martin's letter to Idris presents a model of communitarian spirit and intergenerational care in an almost oversimplified logic: "People speak carelessly of her, because she is old and infirm, as if we must not all, if we are spared, become so; and then, when the young are old themselves, they will think that they ought to be taken care of."[199] In presenting the perspective of Lucy, a servant girl, as equally valid and worthy of Verney's (ill-fated) attempt at saving both her and her mother, the novel foregrounds the necessity of care and humanity over individuality and class consciousness.

Both Juliet and Lucy Martin reinforce the connection between femininity and care, in their dedication to caring for and saving others, but they also serve the narrative function of presenting (distinctly gendered) counter-positions to the general

195 Shelley, 371.

196 Shelley, 370.

197 Albright, "'In the Mean Time, What Did Perdita?': Rhythms and Reversals in Mary Shelley's *The Last Man*," 8–9.

198 Shelley, *The Last Man*, 406.

199 Shelley, 351.

apocalyptic narrative at a point in the novel (chapter 23 to 25) when the path towards death seems imminent and inevitable. Martha, Juliet and Lucy are, moreover, presented as extraordinarily brave and willing to save others (Juliet saves Verney from certain death at the hands of the prophet, Lucy tries everything to save her mother and Martha is focused on saving her whole village). While none of their efforts are ultimately sufficient to stop the apocalypse, they are nevertheless reiterated as ethical examples of good communal leadership, humanity and extraordinary bravery.

The parallel construction of their interspersed narratives serves as a reminder of the novel's complicated structure that juxtaposes Verney's retrospective tale with the stories of these side characters who are – with the exception of Martha – picked up several times only to be discarded along the way. As Albright points out, "[t]hese narrative rhythms mirror the larger structure of the novel within which they are enclosed; we already know from the outset that it is a story of the last man alive on earth and any suspense can only be associated with the intermediate events."[200] Albright connects this temporal structure with Verney's "speculations on fate which occur with increasing frequency during the final chapters of the novel"[201] and argues with Mellor that the "elaborate temporal and narrative" framing of the novel "provides sufficient distance from Mary Shelley for her to deal with the magnitude and complexity of her pain."[202] This reading of the frame narrative as a distancing mechanism is well within the conventional use of frame narratives in the Gothic (employed, not least, in *Frankenstein*) and echoes the sentiments of the frame narrator concerning "human nature" and the "imagination, painter of tempest and earthquake, or, worse, the stormy and ruin-fraught passions of man [which] softened my real sorrow and endless regrets, by clothing these fictitious ones in that ideality, which takes the mortal sting from pain."[203] It does, however, not explicitly account for the prophetic elements connecting the frame with the main narrative level of the novel, nor does it address the complicated gender structure inherent in the framework, prophecies and interspersed narratives.

Gender and Prophecy

As I have argued above, the novels frames add two women's voices from different time periods, and pointing in different temporal directions, to Verney's main narrative level. The fictionalized author visiting the caves in 1818 and commenting on her reconstruction of the Sibylline leaves over "many hours of solitude" adds a level of

200 Albright, "'In the Mean Time, What Did Perdita?': Rhythms and Reversals in Mary Shelley's *The Last Man*," 8–11.

201 Albright, 11.

202 Albright, 12.

203 Shelley, *The Last Man*, 7.

reminiscence of a "time [that] is gone; and, with the selected and matchless companion of my toils, their dearest reward is also lost to me",[204] but also comments on the "imagination and power" inherent in the task of piecing together the narrative.[205] Within the account of this fictionalized Mary Shelley, the Sibyl's leaves are presented as "poetic rhapsodies, and the divine intuition which the Cumean damsel obtained from heaven."[206] While the form of the leaves and the fact that they "were unintelligible in their pristine condition" account for some "distortion and diminution" on the part of the decipherer,[207] the prophetic content of the leaves is never questioned in the frame narrative and its veracity is supported by the discovery of prophecies of "events but lately passed; names now well known, but of modern date;" in "modern dialects, English and Italian" on the "leaves strewed about" in her cave.[208] The fictional persona of the author thus vouches for the accuracy of the Sibyl's prophecies, and, by extension, the prophetic nature of Verney's narrative. This structure is further complicated by a number of prophecies narrated – and ultimately fulfilled – within Verney's own narrative, which, like the frame narrative, draw attention to the novel's negotiation of gender roles, juxtaposing a male-dominated past with a future negotiated, at least in part, by female figures.

There are two major prophecies reported by Verney, and both are related to Lord Raymond (modeled on Byron). The first one is a "presentiment of ill" which hangs over Lionel's sister Perdita, who fears that her husband, Lord Raymond, will abandon her and her daughter Clara for Greece if he fails in his plan to be elected Lord Protector of England.[209] Though he narrowly wins the election, his reign is short, and he ultimately betrays Perdita and returns to Greece to die in the war against the Turks. Perdita's premonition, while at first presented as something of an overreaction, is, thus, justified in the long run, introducing, as Suparna Banerjee argues, with Audrey A. Fisch, an "undercutting of the 'idealization of the male leader' – Perdita's is 'a female voice that first fractures the myth of (male) unity under Raymond'."[210] While Perdita is introduced as possessing a "less perfect" nature than Verney's wife Idris, mainly based on her rough upbringing as an orphan, the "tenderness and happiness" early in her marriage with Lord Raymond evokes an element of personal growth that "improved her temper, and softened her natural reserve. Her understanding was clear and comprehensive, her imagination vivid; she was sincere, generous, and reasonable."[211] Raymond, by comparison, is shown as impulsive,

204 Shelley, 6.

205 Shelley, 7.

206 Shelley, 6.

207 Shelley, 7.

208 Shelley, 5.

209 Shelley, 93.

210 Banerjee, "Beyond Biography," 521.

211 Shelley, *The Last Man*, 91.

possessing violent passions, easily bored and prone to making dangerous decisions, further justifying Perdita's premonitions.

It is one of the novel's ironies that Perdita's agency as a female character only fully emerges in the moment of her death, as she decides to go against her brother's efforts to take her back to England and rather drown herself than be taken away from Raymond's grave near Athens. While Mellor reads Perdita's suicide as embodying "Mary Shelley's recognition that the genderdetermined role of devoted wife within the bourgeois family is inherently suicidal",[212] Banerjee observes that Perdita's devotion to Raymond also highlights her inability as a woman to participate in society by herself and that her attempts to mold herself into an ideal of womanhood is, ultimately, "self-destructive".[213] While Perdita remains devoted to Raymond in the end, their separation leads to a brief moment of understanding, in which she contemplates the necessity of independence:

> He, she thought, can be great and happy without me. Would that I also had a career! Would that I could freight some untried bark with all my hopes, energies, and desires, and launch it forth into the ocean of life – bound for some attainable point, with ambition or pleasure at the helm! [...] But my nerveless hands can neither fell the trees, nor smooth the planks.[214]

It is particularly interesting that Perdita's desire for a "career" of her own, a mode of self-expression that would allow her to live out her "ambition" is expressed in the imagery of both craftsmanship (specifically shipbuilding) and the interaction with the non-human natural world that emerges most prominently in the metaphorical construction of "the ocean of life". Perdita's suicide, in an ocean of death, therefore, does not only signal the loss indicated in her eloquent name, but also the loss of opportunity (for women) that emerges prominently in this passage, signaling the novel's interest in questioning the contemporary gender roles that denied women the opportunity of self-determination.

Perdita forms part of a doomed love triangle, which pairs her with Evadne, another "victim of this genderbased disparity of opportunities",[215] as well as the author of the second prophecy regarding Raymond. Evadne is introduced earlier in the text as a Greek princess and daughter of the Greek ambassador to England, as well as Adrian's secret love interest, who, however, loves Lord Raymond, "the deliverer of Greece, the graceful soldier, who bore in his mien a tinge of all that, peculiar to her native clime, Evadne cherished as most dear."[216] Torn between the love for his wife,

212 Mellor, *Mary Shelley*, 154.

213 Banerjee, "Beyond Biography," 524.

214 Shelley, *The Last Man*, 163.

215 Banerjee, "Beyond Biography," 524.

216 Shelley, *The Last Man*, 45.

Perdita, and care for Evadne, who has returned to England sick and impoverished, but still attempting to secretly make a living as an untutored architect, Raymond stumbles through a series of impulsive decisions, which convince Perdita of his infidelity.[217]

When, shaken by this rift and bored by his political role, Raymond abandons his post as Lord Protector and leaves England to once again join the Greek war of independence as a general, Evadne follows him in secret and joins the army in the disguise of a soldier. As Wang argues, "Evadne's fluid transversals of national, cultural, and gendered borders make her one of the only 'free' people in the novel. Indeed, her numerous veiled disguises allow her to circulate freely among the masculine architectural and military circles of England, Greece, and Turkey."[218] Her spectacular death and dying prophecy further serve to underline her importance in the text. She is discovered by Verney, who has come to Greece with Perdita in search of Raymond, and stumbles upon her near Mount Athos as she is dying from a battle wound. To him, Evadne "sadly lamented her hard fate; that a women, with a woman's heart and sensibility, should be driven by hopeless love and vacant hopes to take up the trade of arms, and suffer beyond the endurance of man, privation, labour, and pain."[219] This reiteration of her womanhood serves to draw attention to several instances of her gender-bending in pursuit of personal and political independence, presenting her as a counter-model of femininity to Clara and Idris, but also lending strength and credulity to her prophecy (or rather curse) directed at Raymond:

> [T]here is the end! there we meet again. Many living deaths have I borne for thee, O Raymond, and now I expire, thy victim! – By my death I purchase thee – lo! the instruments of war, fire, the plague are my servitors. I dared, I conquered them all, till now! I have sold myself to death, with the sole condition that thou shouldst follow me – Fire, and war, and plague, unite for thy destruction – O my Raymond, there is no safety for thee![220]

Evadne's prophecy is carried by angry exclamations and the repetition of Raymond's name, which manifests her claim on his life. In a speech act that is half prophecy and half magical incantation, she aligns her agency with that of the elemental "servitors" of "the instruments of war, fire" and "the plague". As Wang has argued, Evadne functions as a representative of both "nature" and the materialization of the framing

217 As Wang argues, the "love triangle of Raymond, Perdita and Evadne hastily prioritizes the ideal of heterosexual love at the expense of Evadne's emergent androgynous poetics in a situation that closely mirrors Shelley's own turbulent balancing act between love and art." (Wang, "We must live Elsewhere," 249.)

218 Wang, 248.

219 Shelley, *The Last Man*, 181.

220 Shelley, 181.

prophecy in the text, suggesting that her early death does not diminish her function as one of only a few characters who recover from infection thus pointing towards a potential future.[221]

Evadne's words function, in part, as a self-fulfilling prophecy driven by Raymond's guilty conscience, but also by the deep melancholy and lack of enthusiasm he feels after the battle of Rodosto. He tells Verney that Evadne has been "visit[ing] his thoughts" and when told of her prophecy he accepts it: "She has said nothing, but what I knew before – though this is confirmation. Fire, the sword, and plague! They may all be found in yonder city; on my head alone may they fall!"[222] While he continues to pursue the siege of Constantinople in the name of the Greek, he tells Verney that "all I ask of Greece is a grave" because "[t]he last words of Evadne were the seal upon the warrant of my death" and "the shadow of Fate's uplifted hand already darkens me."[223] Raymond's conviction hastens Evadne's prophecy, but it is his recklessness in entering a plague-stricken city abandoned by the Turkish army where he is killed in an explosion, allowing an ambiguous reading of Evadne's words as both prophetic and damning to one of the most controversial characters in the novel.

Verney's search for Raymond in the destroyed city draws on the apocalyptic imagery of Byron's "Darkness", reiterating the story of the faithful dog dying close to his dead and defaced master's body in "this city of the dead",[224] but also echoing Evadne's prophecy of "fire, and war, and plague", when Verney discovers Raymond's body "[i]n that part of the town where the fire had most raged the night before, and which now lay quenched, black and cold."[225] These intertextual and intratextual connections reiterate the function of prophecy in the novel, allowing a reading of the text as a counter-position to Byron's bleak and lonely apocalypse and underlining the function of women in this tale of the last man.

While critics have, as Paley points out, focused on the element of isolation and its autobiographical echoes in Shelley's letters and journals,[226] the novel also presents a counter model to the isolated apocalypse of "Darkness" in allowing Verney to revisit the community and happiness of his past within his narrative and in presenting a number of interspersed narratives that highlight ideas of good leadership, community and humanity.[227] While the novel, thus, does not present a cure for the pandemic or a less isolated position for Verney, it foregrounds the possibility of pursuing

---

221 see Wang, "We Must Live Elsewhere," 241.

222 Shelley, *The Last Man*, 184–85.

223 Shelley, 187–88.

224 Shelley, 207.

225 Shelley, 206.

226 see Paley, "Apocalypse Without Millennium."

227 Laura Fiorentini discusses the ramifications of reading the novel as one of the first post-apocalyptic texts, in spite of the fact that large parts of the novel are set before the apocalyptic

happiness in the community with others as well as in the non-human natural world. The interspersed prophecies of Perdita and Evadne (as well as the interspersed narratives of Martha, Juliet and Lucy Martin) serve as constant reminders of the novel's focus on perspectivity.

The Failure of the Sublime

As the central character and lone figure representing Romantic poetic genius, Verney's relationship with "nature" is explored throughout the text by placing him in a series of almost clichéd scenarios, most prominently the lonely, rustic shepherd boy of the novel's early chapters and the Alpine traveler gazing at the sublime landscape around Mont Blanc. While the former situation is somewhat de-romanticized by an emphasis on both Verney's and Perdita's poverty and lack of guidance, the latter reflects earlier views of the sublime scenery in the context of its usefulness as a refuge, not an elevating experience. Shelley quotes her mother's letters in describing the "seedless rock" as "the bones of the world, waiting to be clothed with every thing necessary to give life and beauty",[228] questioning their usefulness as a refuge from the deadly pandemic:

> We vainly sought the vast and ever moving glaciers of Chamounix, rifts of pendant ice, seas of congealed waters, the leafless groves of tempest-battered pines, dell, mere paths for the loud avalanche, and hill-tops, the resort of thunderstorms. Pestilence reigned paramount even here.[229]

The use of "vainly" early on in this description dampens the effect of the description with a sense of futility, which is underlined by the closing sentence, highlighting the ubiquitous "reign" of "Pestilence". The scenery is imbued with death beyond its own destructiveness, which, consequently, does not function as a sublime spectacle, but as a reminder of a hostile earth ruled over by the plague. The evocation of earth as a "tender mother" in countries where she is wont to "nourish her children" is contrasted with the sharp destructiveness of the mountains, which, as in Percy Shelley's "Mont Blanc", is also associated with a different perspective on time. Higgins analyses the Shelleys' engagement with the Comte de Buffon's theories of deep time[230] and, while these connections emerge in more depth in "Mont Blanc", there are echoes of shifting temporality in Mary Shelley's description of the landscape – most prominently in the suggestion of imminent catastrophe in the "pendant" ice

event, as well as its impact on an emerging narrative tradition. (see Fiorentini, "Mary Shelley's *The Last Man* and Post-Apocalyptic Writing.").

228 Shelley, *The Last Man*, 424.

229 Shelley, 424.

230 see Higgins, *British Romanticism, Climate Change and the Anthropocene – Writing Tambora*, 60ff.

rifts and in the repetitive destructiveness of the elements (e.g. "the tempest-battered pines", "avalanche" and "thunderstorms"). The last humans (Verney, Adrian, Clara and Verney's younger son Evelyn), are nevertheless granted an experience of a curious balance in this largely hostile landscape, as the "[s]ublime grandeur of outward objects soothed our hapless hearts, and were in harmony with our desolation."[231] Harmony, not the elevation of the spirit, is the central theme here and is, therefore, repeated throughout the passage, which describes the burial of the last member of their group (except for Verney, Adrian, Clara and Evelyn) in the "palaces of Nature's own": "This solemn harmony of event and situation regulated our feelings, and gave as it were fitting costume to our last act. Majestic gloom and tragic pomp attended the decease of wretched humanity."[232]

In the ongoing catastrophe, the pre-rational sublime does not fill the experiencing subject with a sense of awe in the face of greatness, but ultimately serves as an emotional outlet for the protagonist's very real physical experience of terror. As in other instances in the text, the situatedness of the narrator suggests a more direct involvement than, for instance in Percy Shelley's "Mont Blanc", which shows a speaker moved by the sublime eternity of the mountains, but also able to transfer his own aesthetic reactions to questions about the "human mind's imaginings" in general.[233] Verney's description, by comparison, reduces the effect of the sublime landscape to "tragic pomp", by connecting it to the momentousness of the specific occasion of burying the last plague victim in a cave on the glacier. While Verney is later moved by the "wild and sublime" landscape,[234] the reduction of the perspective to that of a funeral site draws attention to the situatedness of the experiencing subject and his emotions, as well as challenging the universality of the sublime experience by filtering it through relationality.

As Patricia Cove argues, the translation of the characters' emotions into aesthetic terms such as the sublime "by seeking out terrifying landscapes that mirror their vertiginous encounters with plague, war and political disintegration, can only partly articulate the full 'precision' and 'force' of their physical trauma."[235] For Cove, aesthetic experience functions as a partial outlet for the protagonist's trauma, while the text establishes "the primacy of the body as the site at which identity is established and the location of pain and affliction in the novel."[236] This understanding of aesthetic experience as a supplement does, however, not fully account for Verney's denial of "nature" and the aesthetic experiences it might provide in the face of

231 Shelley, *The Last Man*, 424.

232 Shelley, 425.

233 Shelley, "Mont Blanc," 849.

234 Shelley, *The Last Man*, 425.

235 Cove, "'The Earth's Deep Entrails,'" 20.

236 Cove, 20.

his ultimate loss and loneliness. As in "Darkness" and in "A World Without Water", "lastness" also precludes sublimation in *The Last Man*. The autodiegetic narrative perspective, moreover, centers Verney's emotions and experiences, which are further narrowed down by the disappearance of all other experiencing subjects to the extent that the existence of the world seems to hinge on its continued construction in his narrative perspective at the end of the text.

Where for the speaker in Percy Shelley's "Mont Blanc" "the mind" and "the source of human thought" indicate a universal (or at least transferrable) human subject position from which the sublime can be experienced and described,[237] Mary Shelley's *Last Man* confronts its readers with a multilayered narrative subjectivity that draws attention to the individual experience and its limitations in the absence of a human species that the experience could, ultimately, be transferred to. This distinction is rooted in constructions of and assumptions about gender, as the direct aesthetical and ethical relationship of the experiencing and writing subject and the "everlasting universe of things" only emerges from a dominant subject position whose ultimate capacity for rational thought has never been called into question.[238] The narrative construction of Verney as a reflector of several gendered perspectives and the multilayered construction of his diegetic position in the frame narrative undermine his straightforward gender presentation to an extent that his experience and, consequently, poetic engagement with the sublime is filtered through Othering and, thus, ultimately denied.[239]

Both Browne's "A World Without Water", which I have discussed above, and *The Last Man*, use counterstrategies to allow the texts to engage with the sublime – one of which is a foregrounding or concretization of the speaker's situatedness with regard to the apocalyptic experience, placing the speaker physically in the text as the experiencing subject (for instance Browne's speaker who is standing on a hill). Beyond this spatial component of experience, Verney's situation as the last man also makes it necessary for him to position himself towards other basic concepts, such as time and "nature", as well as the heritage of human culture, which are rapidly losing meaning in the absence of human society. In the final pages of the novel, he first rejects time, throwing away a stick he has been using to keep track of passing days after the loss of Adrian and Clara,[240] then he experiments with first living in "nature" – as in his early days as a shepherd – only to reject the natural world for human

237 Shelley, "Mont Blanc," 845.

238 Shelley, 845.

239 Wang reads Verney as an androgynous figure, arguing that "Shelley employs a shrewdly revisionist strategy that continuously rewrites notions of community and companionship in the variable terms of gender, race, and species." (Wang, "We Must Live Elsewhere," 235.)

240 This action of rejecting time is fraught with meaning by its context in a novel in which, as Ruppert argues, "Shelley overturns artificially kept time and chronological history, thereby undermining two basic props of both conventional English literature and the British world-

culture, which he seeks in Rome. In the absence of other human company, neither strategy is successful, in part because he does not truly feel that he belongs in either space.

Verney's lone decisions, which span several summarized years, are presented from his individual emotional perspective, which challenges Romantic tenets and themes: "No, no, I will not live among the wild scenes of nature, the enemy of all that lives", suggests that Verney sees himself as an outcast even from "nature", trying to escape from his "miserable" human self.[241] While the wonders of human cultural heritage in Rome keep his imagination focused on the past for a while, "the soul wearies of a pauseless flight; and stooping from its wheeling circuits round and round this spot, suddenly it fell ten thousand fathom deep, into the abyss of the present – into self-knowledge – into tenfold sadness."[242] Rome, with all its cultural heritage, symbolizes a flight into the past, which is also linked to Verney's efforts to relive companionship by writing his autobiographical tale. Verney thus sees his position not as universal or even representative of mankind – although he is, in fact, the last representative of his species, but understands and represents himself in relation to generations past and the absence of a companion in the present. This self-recognition as a "solitary being" influences his writing to the extent that it becomes both meaningless and physically inexpressible to him:

> I streak this paper with the tale of what my so named occupations were – while I shape the skeleton of my days – my hand trembles – my heart pants, and my brain refuses to lend expression, or phrase, or idea, by which to image for the veil of unutterable woe that clothed these bare realities.[243]

As Browne's speaker choked by the absence of water in her nightmare, Verney's ability to tell his tale, to write further of his experiences is directly impacted by his position in his post-apocalyptic present, leaving his past as the only refuge from "torturing nature".[244] The novel adds yet another layer of narrative perspective, introducing Verney's narrative as the book he begins at this point, writing "Dedication to the illustrious dead. Shadows, arise, and read your fall! Behold the history of the last man."[245] While his narrative, which he has been commenting on throughout the novel, seems to come full circle at this point, it simultaneously spirals into a new and different beginning suggested by the dedication, which has not, in fact, been

view in the years following the Napoleonic Wars." (Ruppert, "Time and the Sibyl in Mary Shelley's *The Last Man*," 153.)

241 Shelley, *The Last Man*, 460.

242 Shelley, 463.

243 Shelley, 465.

244 Shelley, 466.

245 Shelley, 466.

part of his narrative so far, and the declaration of his intention to "write and leave in this most ancient city, this 'world's sole monument,' a record of these things"[246] – an intention which directly contradicts the novel's frame and its genesis of Verney's narrative. Wang reads this as an expression of the novel's "proliferating trajectories and frequently unwieldy potentialities that continuously revise Shelley's own discussions of gender, agency, and creativity",[247] complicating the novel's discussion of "lastness" and apocalypse.

To some extent, Verney's tale has already ended with the loss of Adrian and Clara, as he is no longer able to interact with the world without their companionship. As he decides to embark on a sea voyage with uncertain destination at the end of his journey, he does so in the knowledge of the possibility of instantaneous death or facing "endless time" as a "lone wanderer",[248] challenging the "spirits of the storm" and "powers of destruction" to receive him and clasp him forever.[249] The novel's ending, thus, does not function as a new beginning, but, as the other apocalyptic texts I have discussed in this chapter, highlights the speaker's/narrator's inability to continue their tale in the face of a catastrophe that can no longer be aestheticized or sublimated within the context of Romantic aesthetics.

Conclusion

In the course of this chapter, I have read three Romantic apocalyptic texts, attempting to frame commonalities and difference in how they envision and aesthetically approach climate catastrophe. I have read Byron's "Darkness" in conjunction with Mary Ann Browne's "A World Without Water" to draw attention to their similar but still distinct formal framing of apocalyptic dream visions. As I argue the poems present different approaches to speaker positionality and the impossibility of framing the end of the world within a Romantic aesthetics of the sublime. While both poems imagine what Paley terms an "apocalypse without millennium"[250] and without even the slightest possibility of religious redemption, they differ considerably in the ways in which they imagine human reactions and interactions. Where in Byron's poem "war [...] did glut himself" and "the meagre by the meagre were devoured",[251] Browne's text shows a "dead, motionless" natural world inhabited by "women [who] sat them down to weep / Over their hopeless pain."[252] The gendered

246 Shelley, 466.

247 Wang, "We Must Live Elsewhere," 239.

248 Shelley, *The Last Man*, 469.

249 Shelley, 468.

250 see Paley, "Apocalypse Without Millennium."

251 Byron, "Darkness," 717.

252 Browne, "A World Without Water," 157.

differences between these reactions become even more interesting if one considers the complete absence of women and children from Byron's poem in which the only feminized entities are the (absent) moon and the universal darkness, and the fact that Browne describes "men [who] had visions dark and deep / Clouding the dizzy brain"[253] – thus echoing the darkness of Byron's title and the visionary imagery of his poem in a manner that might be read as an ironic intertextual reference to his earlier text. As I have argued, Byron's construction of his text as a "dream that was not all a dream"[254] allows him to reclaim poetic authority for his speaker even in the face of universal darkness that consumes everything at the end of the poem. Browne approaches the problem of claiming poetic authority as a woman writer by physically situating and connecting her speaker to the non-human natural world of the apocalypse and by emotionally and materially connecting the different narrative levels of her poem through the physical response of her speaker.

In the third part of this chapter, I have attempted to retrace the apocalyptic aesthetics of the two poems as they reoccur in the complex narrative structure and engagement with aesthetic responses to major catastrophic events in Mary Shelley's *The Last Man*, which openly engages with Byron himself in the character of Lord Raymond. Shelley's novel recontextualizes the apocalyptic imagery introduced in "Darkness" in a framework of community and care that is also reflected in its gendered narrative framework. What emerges from these readings is a sense of shifting Romantic subjectivities in response to fictionalized or imagined apocalyptic events which can no longer be framed within the aesthetics of the sublime. These emerge in the wake of the climate crisis triggered by the eruption of Mount Tambora in 1815, but also reflect political, cultural and religious concerns that complicate a stable sense of "man"'s relationship with "nature". In the texts I have discussed in this chapter, these changes emerge as shifts in aesthetics, poetics and positionality which draw attention to the fault lines of a Romantic poetic subject position that no longer seems entirely tenable in response to the experience of the effects of climate catastrophe in the early anthropocene.

Secular apocalyptic visions in Romantic literature such as the three texts I have discussed in this chapter, thus, challenge a romanticizing aestheticization of "nature" by drawing attention to the situatedness of the perceiving / speaking subject, whose precarious position in the apocalyptic scenario undermines the possibility of a sublime experience. Where, generally speaking, Romanticism reiterates a human subject position that is emotionally involved in and affected by the non-human natural world but constructs the perceiving subject as separate from the perceived object, secular apocalypse serves as a reminder of the constructedness of this sub-

253 Browne, 157.

254 Byron, "Darkness," 716.

ject position and draws attention to the involvement of humans in the non-human natural world.

The question of gender and the cultural construction of gender roles at the beginning of the nineteenth century adds a further dimension to these discussions, by drawing attention to the ways in which aesthetic categories such as the sublime revolve around constructing the perceiving subject as male and, thus, exile the female positionality from aesthetic experience or construct it as an object. As I have shown above with the examples of Browne's "A World Without Water" and Shelley's *The Last Man*, women's texts often foreground this exclusion of marginalized subject positions from aesthetic experience in the dominant philosophical discourse by drawing attention to the necessity of situated and embodied experience as the prerequisite of an aesthetic subject position and by highlighting connectedness to other humans and the non-human natural world as well as the plurality of perspectives in response to the figure of the autonomous self and "his" mastery of "nature". These different responses to apocalypse in Romantic literature resonate with different, frequently also gendered, responses to climate crisis in the current context which I will further explore in the next chapter on "Mushrooms" and in my Conclusion.

Chapter 4: "Mushrooms"

Introduction - Mycological Romanticizations

Mushrooms "appear" in anthropocene discourses. They pop up in a variety of media and cultural engagements including, but by no means limited to news stories, TED talks, popular science books, social media videos and memes, art projects, nature documentaries and ecocritical zombie films. These sell, promote or proclaim the cultural value of nutritional supplements, experimental building materials, alternative burial methods and recipes for anything from puffball-mushroom pizza to beer brewing and how to care for your sourdough starter. While few of these medial representations are straightforward attempts at commercializing or selling mushrooms, they nevertheless rarely seem to focus *on mushrooms*. That is, they rarely see mushrooms and fungi as representations of themselves, focusing instead on all the interesting and astounding things that mushrooms can *do* or the various uses they can be put to. As I will argue in more detail below, mushroom narratives in anthropocene popular culture display a distinct trend of romanticization – most prominently where they are embedded in discourses around anthropogenic climate crisis. Here fungal representations oscillate between glorifications, mystification and exaggerations of scale, presenting mushrooms as magicians, savior figures and panaceas that is, as the response to humanity's most pressing planetary dilemmas – from radiation to oil spills, and from CO^2 to plastic. As mushroom guru Paul Stamets argues, "mushrooms can save the world."[1] In the eyes of his social media disciples, they already have.

This is not a chapter about mushrooms *per se*, but one that traces mushroom discourses, and my arguments are, consequently, not directed at mycology – let alone mushrooms themselves – but at the narratives that popular culture spins around mycological discoveries and the properties and abilities of mushrooms. This is, as I will argue, a question of framing and scale and, perhaps more importantly for my overall argument, a question of the ways in which mushroom discourses in popular culture tend to align themselves with anthropocene progress narratives in that they

1 see Stamets, "TED: 6 Ways Mushrooms Can Save the World."

proclaim easy solutions to large-scale anthropogenic problems without demanding too much effort or change.[2] Good-news mushroom stories allow viewers to externalize problems – anthropogenic climate change and large-scale extinction – because they promise an all-natural, organic and seemingly effortless solution that seems to allow humanity to continue much as before. To put it bluntly, romanticized accounts that promise simple solutions are comfortable, because, if mushrooms can save the world, "we" don't have to do it ourselves.

I will use the term "romanticization" to describe mushroom discourses that use poetic or metaphorical language and imagery to create additional meaning beyond the immediate explanatory function in the context they appear in, such as framing mushrooms as "magicians" or "experts" at specific tasks. In the texts I discuss, these visualizations often create heroic narratives and, thus, serve as an invitation to read mushrooms as savior figures in anthropocene discourses. They establish an emotional connection that obscures more complex entanglements with the non-human natural world. By signaling an alignment of goals between humans and fungi in which the fungi work *for* humanity in "saving the world" they ultimately reinscribe Romantic narratives about "man"'s mastery over "nature" that critical theoretical approaches to the anthropocene have begun to contest.

While a number of recent mushroom texts frame fungi as heroic in the context of climate catastrophe, Stamets' phrase about mushrooms "saving the world", used both in the subtitle of *Mycelium Running* (2005) and in his 2008 TED talk "6 ways mushrooms can save the world",[3] presents the most prominent example of this function as it rhetorically connects individual scientific discoveries and large-scale anthropocene problems. While Stamets phrases these claims as "can" and not "have already", suggesting that some effort might be required, the rhetorical connection the sentence makes seems to overwrite this minor detail, as further medializations foreground the good-news value of possibility over questions of feasibility or execution. As I will argue in more detail below, Stamets' very generalized claims, thus, do not only serve to generate or heighten readers' and listeners' interest in fungi, but they also frame fungi within an ultimately anthropocentric narrative of climate-saviorism that functions analogously to technosolutionism in that it promises the possibility of saving the world through a *deus* (or *fungus*, in this case) *ex machina* solution that does not require fundamental change. Not only is this romanticized framing of fungi as mysterious agents of change who will come to the aid of humanity in the current climate crisis anthropocentric in its instrumentalization of fungal biology. It also reiterates and reinscribes anthropocentric narratives about the separation of humanity and "nature", obscuring less spectacular narratives of necessary change and complex entanglements.

2 Technosolutionism, which I will address in my Conclusion, functions in a similar way.

3 Stamets, "TED: 6 Ways Mushrooms Can Save the World."

My argument here traces these romanticizing effects in different directions to, on the one hand, embed them in the discussion of Romanthropocene connections that I have outlined in previous chapters and, on the other, to juxtapose them with anthropocene theorizations from the context of the feminist critical Posthumanities, which draw on mushroom narratives to counter narratives of "man"'s mastery over "nature". As I will argue, Romantic engagements with mushrooms can roughly be separated into two different strands, based on their understanding of the mushroom as a plant with stable properties that other things can be compared to, or as a mysterious entity that disrupts attempts at easy representation. The second understanding carries over into more recent theorizations of mushrooms, which focus on the disruptive qualities of fungi.

Anna Tsing's critical project, in *The Mushroom at the End of the World*, for instance, draws on mushrooms not as heroic figures but as indicators of the possibility of change, as they resist being framed within the constant cycle of "narratives of progress and ruin", that has dominated the Western cultural understanding of "nature" as a "backdrop and resource for the moral intentionality of Man, which could tame and master" it.[4] Many types of mushrooms – including the Matsutake that Tsing's anthropological project centers on – can not be cultivated, making it necessary for mushroom hunters, buyers and sellers to follow the rhythms of the mushrooms. In Tsing's view, their entangled stories function as an entry into "a mosaic of open-ended assemblages of entangled ways of life, with each further opening into a mosaic of temporal rhythms and spatial arcs."[5] Tsing describes some of these rhythms in *The Mushroom at the End of the World*, but she also draws attention to the ways in which mushrooms can be theorized further, signaling "patchiness" and a rhizomatic open-endedness that challenges narratives of progress. Mushrooms can, thus, also function as an invitation to theorizations of the anthropocene beyond an exclusionary focus on the "anthropos" whose centrality Tsing criticizes, as I have argued in the "Anthropocene" chapter above.

What I propose in this chapter is, consequently, an attempt at disentangling at least a small part of the rhizomatic network of current mushroom discourses to the extent that I can propose a clearer distinction between different approaches to mushrooms in theoretical discussions of the anthropocene and metaphorical framings of mushroom narratives in popular culture.

The first major node of this discursive network traces mushroom metaphors back towards Romanticism, identifying two different usages of figurative language around mushrooms based on different understandings of mushrooms as plants or as mysterious entities that defer clear meanings and categorizations. While the more prominent mushroom similes are based in an understanding of "nature" as a

4 Tsing, *The Mushroom at the End of the World*, vii.

5 Tsing, 4.

largely fixed and stable entity that can be drawn on to express specific meanings, the failure of metaphor around mushrooms signals an understanding of the non-human natural world as a complex entity that resists attempts at linguistic or material mastery. As I will argue, these two versions of the mushroom correspond with an understanding of the relationship between "man" and "nature" that I have traced in the preceding chapter, as well as with recent anthropocene theorizations. As in previous chapters gendered discourse emerges as a complicating category in this context as these different understandings of mushrooms correspond with critical conceptualizations of masculinized and feminized discourses in Romantic literature.

The second node of mushroom discourse I address in this chapter is the creation of heroic metaphors in popular science texts around mushrooms. I argue that, where they favor accounts of mushrooms as savior figures, they invite pop-cultural reframings as narratives of progress or mastery over "nature". These reframings, by extension, elicit a reiteration of the status quo as they defer the necessity of action in the current climate crisis to an externalized savior figure.

The third connected node, that, in part responds to the perseverance and reiteration of these historical heroic narratives concerns theoretical framings of mushrooms in current anthropocene theorizations – most prominently Tsing's work that I have briefly introduced above and her arguments against narratives of progress and ruin. Tsing's argument retraces these progress narratives to the Enlightenment separation of "man" and "nature" that I have addressed in terms of its emergence in Romantic culture in previous chapters and argues, drawing on Ursula Le Guin, that heroic narratives are counterproductive in describing anthropocene exigencies, because they tend to reiterate these ingrained narratives.

The fourth connected node draws on Le Guin's conceptualization of a "Carrier Bag Theory of Fiction" (1986) as a counternarrative to heroic progress narratives, which I then take to a discussion of her short story "Vaster than Empires and More Slow" (1973) as a concrete example of how an unheroic narrative can be modeled on the rhizomatic structures of mycelium.

The concluding part of this chapter attempts to connect these argumentative threads by returning to popular mushroom narratives beyond the heroic in Aliya Whiteley's autobiographical essays in *The Secret Life of Fungi* (2020).

Romantic Mushrooms

> Do you know anything – that is have you ever thought anything about the production of mushrooms? I want to have a method discovered of producing them in great quantities – because they contain more nutriment than any vegetable substance – & appear to need no manure. & besides they are excellent in a hundred ways. The world wants some Epicure to turn Chemist – & give us a scientific book of cookery. I dream of a thousand things which I could do if settled in a house in the country with a garden.[6]

The brief excerpt quoted above from a letter by Robert Southey to his friend the statistician John Rickman, dated 2 June 1802, does not only express Southey's interest in mushrooms in somewhat superlative tones. His phrasing also suggests that he is interested in producing mushrooms "in great quantities", that is, on a commercial or even industrial scale. The seemingly unrelated comments on the need for a "scientific book of cookery" and the "thousand things [he] could do if settled in a house in the country with a garden", support this point, as they connect with Southey's observation that mushrooms are frugal (needing no manure) and useful (in a hundred ways). The letter, thus, rapidly shifts from the mundane to the utopian in their projected uses and the potential (culinary or other) discoveries to be made by some "Epicure [turned] Chemist". Southey's view is both literal and specific, suggesting an understanding of mushrooms as plants, which can be seeded, harvested and, ultimately, domesticated.[7] It is also steeped in generalization, as it does not specify what kind of mushrooms Southey was thinking of growing – apart from the fact that he was ostensibly planning to eat them (they are not only "excellent in a hundred ways", they also "contain more nutriment than any vegetable substance").[8]

---

6 Southey, "681. Robert Southey to John Rickman, 2 June 1802."

7 He is clearly unaware of the fact that many types of mushrooms actually resist this kind of cultivation. Tsing makes this argument about Matsutake mushrooms which can not be cultivated, but require specific conditions to grow in forest climates (see Tsing, *The Mushroom at the End of the World.*).

8 This hyperbolic construction is repeated in current mushroom texts. Whiteley, for instance, argues in much the same way that "there are a million mushroom recipes to try [...]. Versatility might well be the reason for the success of the Common Mushroom – that, and the ease with which they can be grown en masse, industrially, using a process that has been perfected." (Whiteley, *The Secret Life of Fungi*, 19.)

Southey's thinking is rooted in a concept of productivity and the streamlining of processes that is more typical of industrialist efforts than gardening. For him, the production of mushrooms "in great quantities" suggests itself as a form of progress worth exploring, and one that is associated with they idea of production and consumption.[9] Southey's letter suggests that, having just rented a house in Bristol, the poet felt "secluded [...] as in a convent" and expected to "have leisure to do much" away from his London friends and acquaintances.[10] This is an interesting phrasing, as the move to Bristol seems to have occurred because Southey's wife Edith was pregnant and the house actually sat "on the same row" as that of their common friend Charles Danvers, a Bristol wine merchant. Southey's sudden interest in mushrooms may have stemmed from this sense of isolation from society and the resulting leisure time, but it also conveys an understanding of gardening in general and mushrooms in particular as a project of progress that runs counter to the romanticizations of mushrooms in current cultural discourses that I have discussed above. Moreover, it also emerges as a straightforward framing of mushrooms, which, for Southey, can be understood in terms of growing processes and kitchen chemistry.

Southey's letter reiterates a common Romantic understanding of mushrooms as an odd type of plant with distinct properties – most prominently rapid growth and a certain frugal usefulness. This understanding facilitates a use of mushrooms in Romantic literary texts, where they often appear in simile constructions that produce a clear and relatively stable meaning. While rapid growth is often somewhat more visually framed as a "swelling" that occurs overnight, imbuing these mushroom similes with sexualized overtones, this comparison adds a visual similarity but does, in general, not create meaning beyond the idea of sudden appearance and sometimes disappearance. A second Romantic usage, which I will discuss in more detail below understands mushrooms as complex and mysterious, leading to a deferral or even failure of metaphorical meaning, as they resist being framed in these simple comparisons.

A virtual mushroom hunt conducted on the digital humanities hub *Romantic Circles*[11] reveals that, while Southey's letters contain a number of other references to mushrooms, most of them are similes that describe anything from writing tangents (505 and 2881) to his own children's growth – or rather lack thereof (2197) to fairy

9 He is also ill-informed (mushrooms do need a lot of manure if grown on a large scale) and this is the only mention of this project in the letters, suggesting that it might have turned out to be a somewhat whimsical enterprise – a "mushroom" project in a figurative as well as the literal sense.

10 Southey, "681. Robert Southey to John Rickman, 2 June 1802."

11 "Home." *Romantic Circles*. The website is currently undergoing an extensive redesign which will likely lead to malfunctioning links.

stories (550), which Southey had a specific interest in,[12] with the help of mushroom comparisons. While Southey's name is now associated with the naming of the "liberty cap" mushroom in *Omniana*, a collection of notes published in 1812 and co-written with Samuel Taylor Coleridge, this naming seems to be based solely on another simile – a visual association of the mushroom cap with the form of cap worn by French Republicans. As Mike Jay points out, Southey and Coleridge were unaware of the psychoactive properties of the liberty cap mushroom they described and, in spite of accidental discoveries of the mushroom's hallucinogenic effects in the early nineteenth century, the association between the name and the type of hallucinogenic psilocybin mushroom it now commonly refers to was not made until a more systematic study of mycology emerged in the 1870s.[13] The enthusiasm for growing mushrooms in large quantities expressed in Southey's 1802 letter to Rickman, thus, forms an outlier, as it signals a genuine interest in mushrooms, their properties and uses, but it is also a useful framing for Southey's understanding of mushrooms as plants with stable and scientifically definable qualities. This understanding forms the basis for mushroom similes in Southey's letters and poetry, as well as the literary works of many of his contemporaries. Where Romantic texts present mushrooms outside of simile constructions, these appearances are often based on direct observation of well-known qualities, as in Charlotte Smith's "mist-born mushroom" as part of her careful descriptions of "nature" in "The Heath" and "Beachy Head". The most common association with mushrooms in the work of the Romantics – including Southey – is their rapid and stealthy growth, for instance in Southey's 1798 poem to Margaret Hill where he refers to "mushroom feelings, which spring up".[14]

Perhaps the most interesting mushroom reference yielded by my virtual mushroom hunt occurs in Keats' *Endymion*. The poem constructs the incongruous image of "lowing heifers sleeker than / Night-swollen mushrooms" that occurs as part of an enumeration of "great bounty" bestowed by "our lord Endymion" in the speech of a priest calling everyone to attend an offering ceremony to Pan. The comparison, which seems to draw on the sexual overtones mentioned above, is part of a question asking for confirmation: "Are not our lowing heifers sleeker than / Night-swollen mushrooms? Are not our wide plains / Speckled with countless fleeces?"[15] While the mushroom description seems fairly straightforward, Keats' usage of the "night-swollen mushrooms" in a simile construction that seems to signal the opposite meaning of "sleekness" creates interpretational problems. Peter Henning reads this construction as "[c]oncrete, sensuously organic, and yet profoundly artificial –

12 see Jay, "Fungi, Folklore, and Fairyland." For the letters see Southey, "The Collected Letters of Robert Southey."

13 see Jay, "Fungi, Folklore, and Fairyland."

14 Southey, "To Margaret Hill," 139.

15 Keats, *Endymion*, 13.

its two parts pieced incongruously together – the image's poetic power must somehow be located in its very non-functionality."[16] His reading comes to the conclusion that the image creates a "deferral of meaning" which is rooted in "the motif of the mushroom – itself a token of indeterminacy, historically situated between life and death, animal and vegetable."[17] While I would agree with Henning's reading, this understanding of the mushroom as a third category – a kingdom of its own – presupposes mushroom knowledge that was not available to the Romantics – or at least not in its current form. As Nicholas Money argues, "Linnaeus found the fungi very perplexing and classified them with worms in the 1767 edition of his *Systema Naturae*, but the long-standing consideration of fungi as primitive plants survived this zoological interlude."[18] While it is possible that Keats was drawing on Linnaeus' confusion here, the Romantics' understanding of mushrooms seems to have largely situated them in the plant kingdom, considering them somewhat odd, fast-growing plants.

Mellor's framing of feminine and masculine Romantic discourse provides a different critical framing for Keats' use of a mushroom simile that defers rather than clarifies meaning. As Mellor argues, "Keats [...] resists a masculinist construction of the self"[19] on which an understanding of "man"'s mastery over "nature" could be based. Instead, he "imagines the self as unbounded, fluid, decentered, inconsistent" and "define[s] the true poet as one who possesses a self with permeable ego boundaries that exists *only* in its relations with others."[20] This interpretation is, in part, based on a letter to John Reynolds in which he "compares the gaining of wisdom to the weaving of a spider's web"[21] and suggests that "empathically gained and shared experience" would result in a new understanding of humanity: "Humanity instead of being a wide heath of Furze and Briars with here and there a remote Oak or pine, would become a grand democracy of Forest Trees!"[22]

Mellor's reading of Keats' letters constructs him as a writer who anticipates feminist epistemologies and critically engages with "masculinist constructions of the self"[23] that, for the Romantics, also emerge in an understanding of "man"'s mastery over "nature". The metaphors used in the letter to Reynolds that Mellor analyzes in more detail than *Endymion* suggest an understanding of a relationship between humanity and "nature" that resists an easy framing as "mastery" or even a fusing congruity. The image of the "democracy of Forest Trees" suggests flat hierarchies and

16 Henning, "Romantic Fungi and Other Useless Things," 298.
17 Henning, 299.
18 Money, *Fungi – A Very Short Introduction*, 2.
19 Mellor, *Romanticism and Gender*, 175.
20 Mellor, 175.
21 Mellor, 177.
22 Keats quoted in Mellor, 177.
23 Mellor, 175.

a coexistence of positionalities and incongruous ways of creating meaning. While a more detailed analysis of metaphorical constructions in *Endymion* lies out of the range of this chapter, Mellor's critical framing of Keats' epistemological position towards "nature" and knowledge production provides a possible understanding of Keats' use of a mushroom simile that defers rather than clarifies meaning.

A much more detailed search and analysis would be needed to confirm this across a wider range of Romantic texts, but I see two different strands of figurative language around mushrooms emerging from the examples I have been able to trace. The first one is a relatively straightforward simile-construction that is based on an understanding of mushrooms as a type of plant which is clearly defined by certain properties – most prominently its rapid and stealthy growth. In this understanding, mushrooms appear overnight, and this clearly defined cultural meaning can be used to describe other fast-appearing or -disappearing things. This usage is not concerned with mushrooms in themselves but draws on their properties to describe something else in some sort of metaphorical construction. The second usage that I have attempted to outline with the example from *Endymion*, understands mushrooms as complex and mysterious, leading to a deferral or even failure of metaphorical meaning, as they resist being framed in simple comparisons. This understanding aligns itself with non-anthropocentric theories about humanity's relationship with "nature", which are frequently read as "feminine" discourse, as I have suggested by drawing on Mellor above.

Emily Dickinson's poem "The Mushroom is the Elf of Plants" (written in May 1874) presents an example of this second usage of mushroom imagery in an act that Abi Curtis describes as "a moment of interpretive resistance" that is tied to "the enigmatic figure of the mushroom".[24] Dickinson's work moves beyond Romantic paradigms in many ways, but her contradictory poem functions as a conceptual bridge between Romantic and later engagements with mushrooms that reflect a romanticization of the mushroom as an element of fairy imagery,[25] underlining its mythical over its biological properties. The poem harks back to Romantic engagements with the non-human natural world in that it draws on Romantic personifications of "nature", but also shows a resistance against simple framings that is typical of Dickinson's enigmatic perspective. In five stanzas of relatively short lines "The Mushroom is the Elf of Plants" presents complex and frequently contradictory ideas around the mushroom that are expressed in an intricate network of similes and metaphors:

24 Curtis, "Mushrooming: Resistance and Creativity in Sigmund Freud and Emily Dickinson," 29.

25 see Jay, "Fungi, Folklore, and Fairyland."

The Mushroom is the Elf of Plants -
At Evening, it is not
At Morning, in a Truffled Hut
It stop opon a Spot

As if it tarried always
And yet it's whole Career
Is shorter than a Snake's Delay –
And fleeter than a Tare –

'Tis Vegetation's Juggler –
The Germ of Alibi –
Doth like a Bubble antedate
And like a Bubble, hie –

I feel as if the Grass was pleased
To have it intermit –
This surreptitious Scion
Of Summer's circumspect.

Had Nature any supple Face
Or could she one contemn –
Had Nature an Apostate –
That Mushroom – it is Him![26]

As Thomas Johnson argues, Dickinson's poem underwent several revisions in May 1874. In the version listed by Johnson, "Apostate" in the final stanza is replaced with "Iscariot" and "supple" with "Plated" and finally also "outcast", suggesting variations in how Dickinson envisioned the relationship between the mushroom and a potentially personified "nature".[27] At first glance, Dickinson's poem seems deceptively simple in both its structure and the images it presents in comparing the mushroom to other elements of "nature" and finally attempting to describe its relationship to a personified "Nature" in the last stanza. As Curtis argues, however, the poem ultimately positions itself as thoroughly enigmatic, "embracing and celebrating the unfixable nature of language, the transformative energy of the poetic image."[28] The poem's opening line, describing the mushroom as "the Elf of Plants" and the reference to the mushroom's shape as a "Truffled Hut" – along with other personifications in the text – support the general theme of the poem, associating the mushroom with fairy qualities, which, as Jay argues, is typical for a Victorian tradition of mushroom

26 Dickinson, "The Mushroom Is the Elf of Plants," 565–66.

27 see Johnson, "Note on Poem 1298."

28 Curtis, "Mushrooming: Resistance and Creativity in Sigmund Freud and Emily Dickinson," 29.

depictions in literature and popular culture.[29] It is worth noting, however, that there are no elves or fairies in Dickinson's poem, and the elf-connection in the opening line remains entirely metaphorical. Instead, the poem uses this seemingly supernatural quality of the mushroom to explain its "nature" or essence.

Drawing on the Romantic tradition, the poem's first stanza imagines the mushroom as a type of plant, highlighting its ability to grow rapidly and unobservedly. The text uses monosyllabic words rich in plosive consonants (not, Hut, stop, Spot) that mimic minor explosions, which are further highlighted by the capitalizations of words describing the mushroom and its appearance (Hut, Spot). This theme is, however, disrupted by the somewhat contradictory "nature" imagery in the following stanzas and the poem's peculiar meter and rhyme scheme. The second stanza juxtaposes the idea of reluctant or slow growth ("as if it tarried always") with that of the mushroom's rapid appearance and disappearance in the comparison of its whole "Career" or lifespan with fast and fleeting elements of "nature" – the short delay in moving forward caused by a snake's undulating movements and the short-lived existence of a fast-growing but also shallow weed. Neither of these comparisons is straightforward, supporting the reading of the mushroom as somewhat mysterious – "vegetation's juggler" or trickster.

The five stanzas of four lines are written in a mixture of iambic tetra- and trimeters that appears somewhat regular but does not actually present a strictly regular pattern. Nor is there a regular rhyme scheme. Neither of these irregularities is unusual for Dickinson's work, but they are used to create a specific effect in this poem, underlining the deferred meaning around its metaphorical construction of the mushroom. In the first stanza, line 1 and 3 consist of iambic tetrameters, while line 2 and 4 are written in trimeters. This regular meter is, at first glance, supported by a regular rhyme scheme: lines 2, 3 and 4 rhyme, but the first line of the poem does not rhyme with anything else in the text, suggesting that the construction of the mushroom as a "plant" might be an outlier that does not fit with the other interpretations presented in the poem. The second stanza further deviates from the non-pattern emerging in the first: lines 1, 2 and 4 are written in iambic trimeters, only line 3 is written in iambic tetrameter. The rhyme scheme of this stanza relies on two pairs of imperfect rhymes (always and Delay and career and Tare) – suggesting a disruption of a regular system that is similar to the mushroom's surprising interventions into a perceived regularity of "nature". This pattern is, once again, not repeated exactly in the same way in the third stanza. While the stanza follows the same metric pattern as the second stanza, the rhyme scheme is broken up: lines 2 and 4 rhyme (Alibi and hie), but lines 1 and 3 (Juggler and antedate) do not – nor do they rhyme with other lines in the poem. "Juggler" represents one of the central themes of the poem – the mushroom as a trickster figure. "Antedate" seems to point into

29 see Jay, "Fungi, Folklore, and Fairyland."

the future, towards sudden mushroom appearances, which, as the lack of rhyming counterpart suggests, will take place in unforeseen ways.

Stanza 4 offers yet another variation of the metrical pattern, opening on a line written in iambic tetrameter, followed by three lines in iambic trimeter. Lines 2 and 4 are connected by a pair of highly imperfect rhymes (intermit and circumspect), underlining the meaning of the rhyming words and their careful disruption of the text, while line 1 and 3 again do not rhyme, leaving the fourth stanza almost entirely rhymeless. The stanza also introduces a shift in perspective, as it moves from postulating somewhat curious descriptions of mushrooms in the voice of a covert speaker to an overt voice contemplating how the grass might feel about the mushroom appearing among its blades: "I feel as if the Grass was pleased" does not only signal a deviation from the rest of the poem in terms of the speaker's involvement ("I feel"), but also transfers this sudden emotional point of view to an element of "nature", "the Grass", which is not only personified by this implication of an ability to feel but also capitalized and thus highlighted as an entity capable of such emotions. While the voice seems to speak for "the Grass" here, the "I feel" construction seems to relativize this assumption, as this "feeling" could, of course, be incorrect. The stanza thus "intermits" the narrative about mushrooms the poem has been constructing. The capitalization of "Scion" – the representative of the mushroom in this stanza – and "Summer" suggests that these three entities seem to exist on an equal footing in the stanza, connected by an emotional bond that the speaker can "feel" but not perceive directly. The stanza presents the most straightforward description of the mushroom as a careful, perhaps not fully trustworthy sign of summer, as well as the suggestion that the relationship between the mushroom and the grass might be interpreted within an emotional range (feeling pleased) usually attributed to humans. Stanza 4, thus, emerges as the most "Romantic" part of the poem, highlighting an emotional involvement of the speaker along with a personification of elements of the non-human natural world that is typical of Romantic poetry. The emerging image is, nevertheless, contradictory, as the somewhat negatively connoted "intermit" and the "as if" suggest that the speaker might be entirely wrong in their interpretation. Johnson's discussion of earlier versions suggests that this verse was added at a later time, which might explain the slight difference in tone.[30]

The final stanza continues the personification and capitalization of "Nature", seemingly proclaiming the mushroom as "Nature's Apostate" (or "Iscariot" in the earlier version), but simultaneously relativizing this proclamation by putting it in the context of an unfulfilled condition: "Had Nature an Apostate" suggest that "Nature" does not have an Apostate, nor could she "contemn" one if she had one.[31] The replacement of "supple" with "outcast" and "Apostate" with "Iscariot" in Johnson's ver-

30 see Johnson, "Note on Poem 1298."

31 Dickinson, "The Mushroom Is the Elf of Plants," 566.

sion suggests stronger religious connotations to the stanza, as "Nature" becomes a godlike figure whose capacity to outcast and condemn a Judas figure would then be relativized by the unfulfilled condition. The published version's use of "supple" is more enigmatic, especially as "supple" now seems to modify "Nature"'s face, while "outcast face" referenced the Judas figure in the penultimate line. The suppleness of "Nature"'s face also suggests that this, if it existed, would be an embodied, rather than a mental feature, which modifies but does not clarify the poem's understanding of "nature". The stanza's inherent contradictions suggest that "Nature" resists universalization or simple representation by one "face" or even one adversary. This resistance to categorization in the final stanza refers the poem back to the initially proclaimed status of the mushroom as "the Elf of Plants" – supporting its equally ephemeral appearances and prefiguring an understanding of fungi as outside of a preexisting, but already complex system of "nature".

The somewhat tenuous structure of this last stanza supports the fragile status of the mushroom, as it repeats the metric and rhyming pattern of the third, introducing yet another slight variation: the stanza opens on a line with four stressed syllables, connecting lines 2 and 4 with a pair of imperfect rhymes (contemn and Him). Lines 1 and 3, however, end on "Face" and "Apostate", which function as eye rhymes, thus, continuing the pattern of slight changes in rhyme scheme or meter throughout the poem. This seemingly chaotic structure, which is a variation on Dickinson's preferred pattern of the common meter, supports the theme of mushrooms as they emerge in the poem, as "Vegetations Juggler[s]" appearing suddenly, like "bubbles" in unexpected places.[32]

The poem's shift at the beginning of the fourth stanza from a covert speaker, postulating poetic facts about mushrooms, to an overt speaker, describing an emotional understanding of the mushroom's position in the natural world, is indicative of its roots in Romanticism, but it also suggests an awareness of positionality. While the mushroom is the main subject of the poem and the speaker mostly disappears behind the description of its qualities, there is a slight break in this pattern at the beginning of the fourth stanza which postulates an "I", before the poem shifts its focus back to the non-human natural world with the discussion of "Summer" and "Nature" before ending on the proclamation that "it is Him!" The capitalization of the pronoun

32 Dickinson, 566. The reference to "bubbles" not only functions as natural imagery, but also evokes a line from Shakespeare's Macbeth, comparing the witches to bubbles of the earth (instead of the water), because of their ability to vanish into thin air. This intertextual connection supports the poem's construction of mushrooms as belonging to the supernatural as well as the natural world, underlining the theme of location also evoked by the line "the Germ of Alibi" and the rhyming "hie". The mushroom appears unexpectedly in both a temporal and a spatial context.

and the exclamation mark suggest a definite statement, that might have carried additional religious connotations in Johnson's alternative version. The proclamation is, however, negated in advance by the above-mentioned "[h]ad Nature..." construction of the whole stanza.

The result of this construction is somewhat ambiguous as the seemingly straightforward statement in the last line is relativized while the proclamation still stands –"it is Him!" – not "it would be him". As a candidate for the position of Nature's "apostate" the mushroom, moreover, takes up a complicated stance outside of or pitted against the all-encompassing entity of "Nature" with a capital "N". As Curtis argues, this resistance to clear, unambiguous meaning is also enacted in the poem's "morphing of language which goes beyond any fidelity to the natural world [it] seems, on the surface, to describe."[33] In this, Dickinson's "poem illuminates the peculiarly apt nature of the mushroom image as a figure for the problems of metaphor."[34] While the mushroom is described with a plethora of metaphorical constructions throughout the poem – some of which the poem negates, as I have argued – very few of these metaphors and similes are productive of meaning *for* the mushroom. Instead, they produce a complex patchwork of seemingly unrelated images from which the figure of the mushroom emerges as a "juggler" or "elf", who can not be fixed in one definite meaning, temporality or location, but, instead, draws on a plethora of relations to other elements of the non-human natural world.

This poetic construction of the mushroom contradicts the straightforward understanding of the mushroom as a plant that other things or people might be compared to, which I have outlined earlier in this chapter. Instead, Dickinson's poem functions as an example of a second, more enigmatic version of poetic constructions of the mushroom which understands it as a figure of deferred meaning that can be used to criticize the construction of simplistic "nature" metaphors or similes. In this, both the mushroom and "nature" emerge as entities that resist master narratives and assimilations into a framework of poetic construction that might facilitate the proclamation of an "autonomous self" in response and connection to "nature". The mushroom does not emerge as an heroic figure in "The Mushroom is the Elf of Plants", but is constructed as a somewhat mischievous juggler resisting stable interpretations and clear categorizations.

While I will return to literary mushrooms later in this chapter, I will outline the more recent instrumentalization of mushrooms as hero figures in popular science narratives in the next part of this chapter. As I will argue, mushroom narratives in anthropocene popular culture display a distinct trend of romanticization, which leaves them open to being instrumentalized as heroic tales of saviorism, reiterating a comfortable distance between humanity and "nature" in the current situation

33 Curtis, "Mushrooming: Resistance and Creativity in Sigmund Freud and Emily Dickinson," 37.

34 Curtis, 30.

of acute climate crisis. These romanticized narratives do not only align themselves with progress narratives promising easy solutions, but they also reflect an understanding of "man"'s mastery over "nature" in the figure of the mushroom "savior" and his inventive mycologist sidekick who will save the world.

Romanticizing Mushrooms

As Giovanni Aloi argues in the editorial of a recent issue of the journal *Antennae* on "Mycologies" (2022), "[o]ftentimes, our ability to think of something is not simply defined by the limitations of our perceptual boundaries but by the limitations that institutions implicitly place upon us – what they allow us to think and say."[35] Textual conventions can function in this way, imposing boundaries and limitations on what counts as truth, as a story or as a perspective,[36] as can disciplinary or genre boundaries. As Tsing argues, "progress stories" have a tendency to eclipse other ways of knowing the world. Instead, her mushroom-centered approach "sketches open-ended assemblages of entangled ways of life, as these coalesce in coordination across many kinds of temporal rhythms."[37] There is a fundamental difference between Tsing's theorization of mushrooms and the popular mycological narratives I discuss in this chapter, in that Tsing sees mushrooms as an invitation for a change in perspective – not just on mushrooms, but on the ways in which engagements with mushrooms may lead to a non-progress-oriented way of understanding and interacting with the world.

Aloi's argument, similarly, draws attention to and sees artistic engagements with mushrooms as a way of arguing against the "patriarchal matrix that anchors both western philosophy and science to a bedrock of methodological dehumanising approaches", suggesting that an integration of art into these discussions allows a shift towards less biased approaches, as "today's outlook on the non-human in academia and art is grounded on empathy and imagination."[38] To underline this shift, Aloi proclaims a recent "mycological turn" in the humanities, using the metaphorical language of fungal networks:

> Mycorrhizal networks are the perfect metaphorical incarnation of the webs of knowledge that the different strands of academic research are generating today – open-ended and poietic support networks that gather and convey information

35 Aloi, "Editorial," 10.

36 see Le Guin, *The Carrier Bag Theory of Fiction*.

37 Tsing, *The Mushroom at the End of the World*, viii.

38 Aloi, "Editorial," 11.

> across multiple individuals, shared communities, stratified and often invisible underground agents.[39]

While I would agree with Aloi's sense of a mycological turn – especially in the feminist critical Posthumanities – the obvious metaphorical parallel he chooses to describe this turn might result in a somewhat misleading understanding of what it encompasses. There are, of course, multiple connections between the theoretical discussion of mushrooms and the ways mushrooms are used to make art and explore cycles of life that bring together humans and mushrooms in unexpected ways. Haraway's thoughts on composting in *Staying with the Trouble* (2016),[40] for instance revolve around similar human decomposition problems of the anthropocene as those addressed in more limited hands-on or artistic ways by the inventors of mushroom coffins[41] or Jae Rhim Lee's mushroom burial suit art project[42] and her conceptualization of "decompiculture".[43] I would, however, suggest that these critical and artistic approaches engage with anthropocene problems from different discursive positions and with very different intentions to those evinced by mushroom narratives in popular culture which draw on the mushroom as a comforting savior figure instead of engaging with its disruptive potential. To connect critical theorizations and popular romanticizations within the same metaphor – perfect as it sounds – runs the risk of obscuring these distinctions and obfuscating critical meaning to the extent that the theoretical and artistic engagements and biological discourses can also be framed as just another vaguely interesting mushroom story to scroll by on social media, feeding the general interest in all things "fungal".

In this part of the chapter, I will juxtapose mushroom discourses from different contexts – including biology, anthropocene theory and popular culture, arguing that mushroom narratives in anthropocene popular culture display a distinct trend of romanticization – most prominently where they are embedded in discourses around anthropogenic climate crisis. I locate these romanticizations in heroic narratives that emerge from the texts themselves, but also trace them in reframings of these approaches and the ideas they present in various medializations in popular culture.

I will use the term "romanticization" to describe these heroic savior narratives that instrumentalize mushrooms as heroic figures to save humanity – stretching the meaning of the term beyond its common use as a synonym for "idealize" or "sentimentalize" to describe a process that entails the use of poetic or metaphorical language and imagery in non-fictional texts with a predominantly pragmatic function

39 Aloi, 11.

40 see Haraway, Staying with the Trouble, 159.

41 Page and Deputato, "Dying to Be Green."

42 Lee, "TED: My Mushroom Burial Suit."

43 Lee.

(to convey knowledge or to inform readers about the biology, properties and capacities of mushrooms in a manner that may be entertaining but is, predominantly, instructive). I am particularly interested in metaphorical constructions that create additional meaning beyond the immediate explanatory function in the context they appear in, such as framing mushrooms as "magicians" or "experts" at specific tasks. While metaphorical language and imagery are, of course, functions of everyday language and not limited to poetic or fictional texts, I argue that this particular use of romanticizing figurative language carries additional functions that move beyond simple visualizations of mycological intricacies. In the texts I discuss, romanticizing images create heroic narratives and, thus, serve as an invitation to read mushrooms as savior figures in anthropocene discourses. I argue that this usage creates entry points which leave the narratives open to instrumentalization in calming or comforting narratives in current popular culture, which reiterate rather than disrupt the understanding of humanity's relationship to the non-human natural world.

The examples I will discuss in this chapter situate themselves on a spectrum between the two poles of popular mycology and mushroom theory that align themselves with two contrasting narratives about the relationship between humanity and "nature" in the anthropocene. Theoretical approaches such as Tsing's *The Mushroom at the End of the World* present pressing arguments for a necessary reframing of this relationship to resist progress narratives, while mushroom narratives in popular culture drawing on the work of popular mycologists such as Stamets have begun to imagine mushrooms as savior figures that will single-handedly solve humanity's problems in the anthropocene. Stamets, as I will argue, invites these framings in his romanticization of mushrooms, while other mycologists – Merlin Sheldrake for instance – do not necessarily align themselves with romanticizing mushroom discourses in the same way. To clarify my argument, I will nevertheless attempt to present these two approaches to framing mushrooms in the anthropocene as opposing strategies, beginning with the ways in which they emerge in popular mycological texts.

Both Stamets' and Sheldrake's work unites mycological niche concerns and popular culture, oscillating between romanticizations of mushrooms as the ultimate savior figures and a more multifaceted approach that combines biology and art. Stamets is, perhaps, the most famous pop-cultural figure in this context. His mycological public persona led to the naming of an eponymous *Star Trek* character who develops a "spore drive" that allows starships to jump from one point in space to another, bridging huge distances by using a mycelial network.[44] This utopian framing in popular culture echoes Stamets' own discussion of mushrooms as possessing the ability to "save the world" – a narrative which does not only resonate with *Star Trek*'s utopian worldview, but also creates further entry points into Stamets' work on mushrooms, as *Star Trek* fans began to engage with his popular presentations of

44 *Star Trek Discovery*: "The Butcher's Knife Cares Not for the Lambs Cry."

mushroom narratives. His TED talk on "6 ways mushrooms can save the world"[45] is among the top 10 of viewer favorites on the TED website and has currently more than eight million views on Youtube, suggesting that his popular persona has now begun to overshadow his actual mycological work.

In the foreword of Stamets' book *Mycelium Running*, Andrew Weil suggests that mushrooms "may turn out to be important keys to both human health and planetary health" based on their "indispensable role in recycling organic matter."[46] As the foreword argues, the book circles around "Paul's Big Ideas",[47] gesturing towards both Stamets' own fascination with mushrooms and the popular preoccupation with an idea of mycological genius that can, perhaps, best be observed in Stamets' public appearances. What I am interested in here is not so much the plausibility of Stamets' mycological claims, for example that "[w]ithout fungi, all ecosystems would fail",[48] but the ways in which they are framed in popular culture discourses that seem to romanticize both the mushrooms themselves and the mycologist. I do not mean to suggest that mushrooms might not be useful in mopping up some of the mess humans have created, but rather that to frame them as savior figures buys into the kind of progress narrative driving the anthropocene.

I argue that the superlatives used in describing and visualizing mushroom properties for the larger public provide an invitation for such romanticizations. Stamets, for instance, while fully invested in an educational project around fungi and their uses to humanity and the planet, praises mycelial structures as "mycomagicians"[49] and claims that "mushrooms emerge as environmental guardians in a time critical to our mutual evolutionary survival."[50] The association of mushrooms with magic is not unique to Stamets' work. Biologist and mushroom expert Sheldrake claims that fungi are "metabolic wizards", using this framing to explain that mushrooms "can break down some of the most stubborn substances on the planet, from lignin, wood's toughest component, to rock, crude oil, polyurethane plastics and the explosive TNT."[51] While both Sheldrake's and Stamets' projects are, ultimately, focused on conveying information and drawing attention to the largely unexplored world of fungi, their common use of the pre-rational language of mushrooms as "magic" seems to be rooted in pop-cultural framings of mushrooms as associated with both the imagination and hallucinogens. Beyond the medical properties of various mushroom species, the ideas these framings evoke are based on myth as well as literary

45 Stamets, "TED: 6 Ways Mushrooms Can Save the World."
46 Stamets, *Mycelium Running*, viii.
47 Stamets, viii.
48 Stamets, 1.
49 Stamets, 1.
50 Stamets, 1.
51 Sheldrake, *Entangled Life*, 5.

and other medial representations of mushrooms, for instance various adaptations and pop-cultural engagements with Lewis Carroll's *Alice's Adventures in Wonderland* (1865) – presumably impacting the presentation of mushrooms as mythical in popular science texts.

As Whiteley, argues, this type of framing leads to a disengagement from the underlying problem as well as the mushrooms themselves, as a feelgood story of plastic-eating or other pollution-diminishing fungi suggests that climate crisis has now been solved: when these stories are disseminated "in news stories or on social media [a] few broad strokes of the pen paint a picture of a better world, achieved with the minimum of fuss, and the fungus itself barely gets a mention. It falls into that category of things we don't need to understand."[52]

While Stamets' public persona has taken on a life of its own, further driving his mycological work and his commercial enterprises in selling mushroom supplements,[53] Sheldrake is first and foremost a biologist and his *Entangled Life* (2020) presents a detailed and sometimes personal account of mushroom biology. The presentation of his work on his website, however, also features a video of oyster mushrooms sprouting from a copy of *Entangled Life*, set to music featuring the recordings of the fungus devouring the book[54] – as well as a video of Sheldrake cooking and eating the mushrooms sprouting from the book.[55] Sheldrake's *Entangled Life* also opens with his own wonder at discovering the minutiae of mushroom life through his work as a field biologist in the tropical forests of Panama in terms that engage with the limits of his own and others' perceptions.[56] The images of ink caps and other mushrooms in *Entangled Life* are "drawn with ink made from shaggy ink cap mushrooms",[57] creating a sense of circularity and attention to mediality. Entanglements with fungi seem to be at the forefront of these mycological texts, but both foreground the positionality and perspective of the writer in these interactions, shaping their view of mushrooms to an anthropocentric perspective, and both suggest an entanglement of art, biology and sustainability that seems to be mainly a marketing strategy, but uses narrative elements that are particularly prominent and foregrounded in these recent engagements with mycology. It is, thus, not only Stamets who seems to engage with mushrooms on a level that goes beyond the mere observation into personal and emotional territory, but his presentations of mushrooms often sound more similar to a sales pitch than a biology book.

---

52 Whiteley, *The Secret Life of Fungi*, 35.

53 see "Host Defense® Mushrooms."

54 Sheldrake, "Oyster Mushrooms Sprouting from Merlin Sheldrake's Book, Entangled Life."

55 Sheldrake, "Merlin Sheldrake – Videos."

56 Sheldrake, *Entangled Life*, 12.

57 Sheldrake, 8.

While Sheldrake praises the survival skills of "radiation-resistant organisms" that "may help to clean up nuclear waste sites",[58] Stamets, reading the biological world as political, suggests that a "United Organization of Organisms" would certainly vote humans off the planet. When humans "irresponsibly exploit the Earth, disease, famine, and ecological collapse result", so we "face the possibility of being rejected by the biosphere as a virulent organism."[59] These arguments align Stamets himself and, potentially, other humans acting "as a responsible species"[60] with the ecological world. His stance romanticizes mushrooms not only as "our fungal friends"[61] but also as the "neurological network of nature" that has "the long-term health of the host environment in mind."[62] Drawing on James Lovelock and Lynn Margulis' work, Stamets sees "mycelium as the living network that manifests the natural intelligence imagined by Gaia theorists",[63] arguing not only for "fungal consciousness",[64] but imbuing it with a sense of drive and agency that resonates with personifications of "nature" in the Romantic period. While these narratives seem to highlight mushroom biology, their framing in popular culture has led to an understanding of mushrooms as the ultimate savior figure which will save humans from our own actions and even from ourselves. Stamets' mushroom supplements are branded as "Host Defense", suggesting a close cooperation between human and fungus in which the "fungi perfecti" are framed as a biological army defending the human host.[65]

Other stories also hinge on superlatives, romanticizing mushrooms as saviors of the world. Stamets' *Mycelium Running*, while overall striving to convey useful information about fungi, does so in a language that often verges on the apocalyptic:

> About 250 million years ago, at the boundary of the Permian and Triassic periods, a catastrophe wiped out 90 percent of the Earth's species when, according to some scientists, a meteorite struck. Tidal waves, lava flows, hot gases, and winds of more than a thousand miles per hour scourged the planet. The Earth darkened under a dust cloud of airborne debris, causing massive extinctions of plants and animals. Fungi inherited the earth, surging to recycle the postcataclysmic debris fields.[66]

---

58 Sheldrake, 5.

59 Stamets, *Mycelium Running*, 1.

60 Stamets, 1.

61 Stamets, 1.

62 Stamets, 2.

63 Stamets, 4.

64 Stamets, 7.

65 see "Host Defense® Mushrooms."

66 Stamets, *Mycelium Running*, 3.

Phrases such as "[t]he Earth darkened" and "[f]ungi inherited the earth" draw on biblical language and imagery without making explicit religious references, in a way that is very similar to Byron's vision of the end of the world in "Darkness" (discussed in detail in the chapter on "Apocalypse" above). As Byron, Stamets draws on apocalyptic language to evoke specific images and emotions, but he also replaces humanity with fungi in a post-apocalyptic tale of recycling and clearing the "debris fields" that suggests an employment of fungi in a larger project of making the world "habitable" – not for themselves, but for humans, presumably.

Sheldrake, whose approach to fungal narratives is overall more reticent and less prone to superlatives, describes Stamets' narrative style as

> straight out of an American blockbuster movie. Many of his accounts feature fungal heroes, poised to save the planet from almost certain doom. *Viral storms of unprecedented proportions threaten global food security. Critical pollinators struggle on under grave threat from the virus-bearing parasites, poised to inflict global famine. The future of the world hangs in the balance. But wait. Is that ...? Yes! Once again fungi come to the rescue with the help of their human accomplice, Stamets.*[67]

While this is an overall accurate, if slightly exaggerated, description of both Stamets' narrative style and his public appearances, Sheldrake's account of Stamets' way of presenting mushrooms also draws attention to the kind of stories that Stamets weaves around the abilities of mushrooms and the ways they might be helpful in saving the world. A narrative of "fungi com[ing] to the rescue" does not only draw on a specific narrative pattern of heroic action, but it also reiterates an understanding of the singular quality and ingenuity of the hero whose presence overshadows other narratives and hampers an understanding of other, less heroic actions that might lead to large-scale change. What Sheldrake does not comment on in his framing of Stamets' heroic narrative is the impact such narratives have in the anthropocene, as they reiterate stories of progress and the achievements of singular heroic figures which will save the world from anthropogenic climate crisis.

Tsing, drawing on Ursula Le Guin's "Carrier Bag Theory of Fiction" suggests that the problem lies with the kind of story that is being told over and over again:

> [S]tories of hunting and killing have allowed readers to imagine that individual heroism is the point of a story. Instead, she [Le Guin] proposes that storytelling might pick up diverse things of meaning and value and gather them together, like a forager rather than a hunter waiting for the big kill. In this kind of storytelling, stories should never end, but rather lead to further stories.[68]

---

67 Sheldrake, *Entangled Life*, 222.

68 Tsing, *The Mushroom at the End of the World*, 287.

While I will discuss Le Guin's poetics of non-heroic storytelling in more detail below, I would like to briefly draw attention to the implications of this understanding of different modes of storytelling for Stamets' approach to presenting mycological ideas here. Tsing draws on Le Guin to suggest a reframing of the stories we tell about the world to move away from heroic narratives towards "patchiness" and "precarity" and the ways in which entanglements of various kinds might help us understand how we can engage in new ways of communal survival.[69] This understanding of storytelling as a form of communal and continuous weaving of stories about the world creates openings and provides possibilities for engagements that the hero narrative precludes by staging the hero as the only person in charge of saving the day.

The hero narrative is, thus, the least engaging way of telling stories about mushrooms, because it hampers engagement and suggests that everything is going to be fine, as long as we believe in the powers of the mushroom savior figure. This is fundamentally counterproductive, because the adoption of mycoremediation strategies relies, in many ways, on communal engagement, not a sense of already having solved the problem once the heroic mushroom figure has been identified. By framing mitigating strategies as heroic tales of lone heroes coming up with singular solutions to the monolithic problem of saving the world, everyone else is soothed into the belief that they can ignore the problem and rely on the hero figure for its solution. This kind of narrative is, thus, not only not *conducive* to actual change on all levels – it stands in the way of even imagining its possibility. The romanticization of mushrooms in popular mycological narratives functions as a contributing factor in this context, as it provides readers and audiences with a pre-modeled emotional reaction that feeds the expectation of an easy, non-invasive solution. In this sense, the presentation of mushrooms as the solution to various ecological problems (as it is frequently employed by radical mycologists such as Stamets and Peter McCoy), functions as a soothing but ultimately fallacious reminder that some radical solution will certainly pop-up and "solve" the problem of anthropogenic climate change once and for all.

Such messages of reassurance are a common denominator in a number of texts and films presenting fungi for popular audiences, including the documentaries *Fantastic Fungi* (2019), which also showcases Stamets' work, and *The Kingdom – How Fungi Made Our World* (2018). Both set spectacular images of fungi to dramatic soundtracks, trying to establish connections to audiences' lives, by suggesting a sense of entanglement between humans and fungi "from your first breath to your last."[70] These arguments hinge on an emotional connection, that is, perhaps, necessary for audiences to maintain an interest in slime molds and underground hyphae, but which, ultimately, serves to reiterate an anthropocentric perspective and a heroic narrative about mushrooms.

---

69 see Tsing, *The Mushroom at the End of the World.*

70 *Fantastic Fungi.*

The problem also hinges on the question of who is assumed to possess agency and in what ways. As Michael J. Hathaway notes, the "major transformations" brought about by fungi over hundreds of millions of years "were not singular, discrete events" and would, therefore, hardly register as "agency" in the human sense. "Thus, while I began to recognize that fungi were indeed 'world-changing,' I still wondered if it was possible to understand their actions as 'world-making.'"[71] While Hathaway acknowledges multiple ways in which fungi can, indeed, be described as world-making, none of them quite suggest themselves as heroic savior narratives. To see mushrooms as able to change the world for the better, as Stamets and McCoy argue, however, does not only rely on a sense of fungal agency, but this view also rests on the assumption of fungal intentionality being aligned with human assumptions of what a "saved" world would look like. My point here is not to discredit radical mycology as a means of finding direly needed solutions of mitigating the impact of climate crisis, but rather to draw attention to the kinds of stories that are being told about these discoveries and what a necessary change in perspective might entail.

Compared to Stamets' "big ideas", Sheldrake's work is more aware of how narratives about mushrooms are framed. In his discussion of the work of radical mycologists like Stamets and McCoy, he draws attention to the limitations imposed on plans of environmental remediation based on fungi, especially on the necessity of "a big institutional boost" to put such plans in action. While there are, of course, fundamental problems in the belief that funding for big systemic change might come from within the system that needs to change, Sheldrake also sees a problem in small-scale engagement. As he argues, he "found it hard to imagine that an army of grassroots hobbyists, no matter how dedicated, could be equipped or credible enough to move things forward."[72] Sheldrake's arguments highlight the ways in which mycologists – even those aware of or working within a framework of radical mycology can only imagine "mycoremediation" within existing structures and systems.[73] As Sheldrake argues with regard to Stamets' involvement in the creation of *Star Trek – Discovery*,

> [o]ne of the problems (the real) Stamets hoped to address by collaborating with the *Star Trek* team is the neglected state of mycology. Art imitates life and life imitates art. Fictional astromycologist heroes might be able to shape the nonfictional future of fungal knowledge by inspiring a generation of young people to get excited by fungi.[74]

71 Hathaway, *What a Mushroom Lives For*, 51.

72 Sheldrake, *Entangled Life*, 207.

73 Sheldrake, 208.

74 Sheldrake, 220.

While there is nothing wrong with "getting excited" about fungal knowledge, Stamets' heroic mushroom narratives may, as I have argued above, create a sense of comfort instead of the desired environmental engagement as they align themselves with Romantic narratives of "man"'s mastery over "nature". I will return to this argument and (eco-)feminist counterarguments that engage with the necessity to think beyond narratives of progress below with regard to Anna Ting's and Ursula Le Guin's work.

Mycelial Entanglements and the Arts of Noticing

> These organisms make questions of our categories and thinking about them makes the world look different. It was my growing delight in their power to do so that led me to write this book. I have tried to find ways to enjoy the ambiguities that fungi present, but it's not always easy to be comfortable in the space created by open questions.[75]

In *Entangled Life* Merlin Sheldrake describes not only his fascination but also his frustration with fungi, arguing that "it's not always easy to be comfortable in the space created by open questions." As Sheldrake shows in biological detail, mushrooms "make questions of our categories", resisting easy categorizations or demanding entirely new systems on an unprecedented scale. While this quality creates discomfort for biologists whose job encompasses the systematization of mushroom knowledge and research, it is precisely this ambiguity and multifaceted resistance to categorization that makes the mushroom a particularly interesting entry point for theoretical reframings of humanity's relationship with "nature" in the anthropocene.

In his introduction to *Entangled Lives*, Sheldrake draws attention to the different ways in which lab biologists and field biologists interact with "the pieces of life they study". While lab biologists are – or at least see themselves as – "in charge" of the object of their study and their own lives are "lived outside the flasks that contain their subject matter", field biologists neither "have so much control", nor do they exist outside and separate of their subject of study: "The world is the flask and they're inside it", as Sheldrake argues.[76] Sheldrake clearly makes this distinction to draw attention to the immersiveness and proneness to accidents and failures more common

75 Sheldrake, 15.

76 Sheldrake, 13.

in field work, but his description also draws attention to the different ways in which positionality is framed in various engagements with fungi. Scientific (that is, in the cases Sheldrake describes, biological) engagements with fungi are imbued with a habitus of objectivity and control, especially were they take place under controlled conditions in the lab, that seems to go unquestioned until it is missing. Sheldrake is aware of the ways in which "anthropocentric definitions" shape what or who is seen as intelligent and that these definitions usually exclude "slime molds, fungi and plants",[77] but as I have argued above, popular mycology still often falls into the trap of presenting mushroom biology as a heroic narrative, staging mushrooms as savior figures who will save the world from the effects of anthropogenic climate crisis.

In this part of the chapter, I turn to anthropocene theorizations of mushrooms – most prominently Tsing's work in *The Mushroom at the End of the World* – not only to present a counterapproach to the popular romanticized mycological narratives I have discussed above, but also to present a reframing of mushrooms as a disruptive element within a larger theoretical context that critiques heroic progress narratives in the anthropocene. As Tsing argues, it makes sense to pay attention to the stories that emerge around people's interactions with mushrooms, precisely because these are so often based on chance encounters and rarely part of grand schemes of progress.[78] In this, Tsing's work aligns itself with Haraway's feminist argument about the role of situated knowledges[79] in the sense that the narratives imagined are in no way universal or heroic, but individual and localized, allowing for a rethinking of systemic inequalities and power structures.

There is a certain urgency to this shift in perspective because, as Tsing argues, the progress narratives which drive climate catastrophe do not offer viable solutions of how to mitigate its effects:

> Industrial transformation turned out to be a bubble of promise followed by lost livelihoods and damaged landscapes. And yet: such documents are not enough. If we end the story with decay, we abandon all hope – or turn our attention to other sites of promise and ruin, promise and ruin.[80]

Instead of turning to mushrooms as savior figures that might get us out of this predicament, Tsing explicitly challenges progress narratives based on their ten-

---

77 Sheldrake, 17.

78 Sheldrake's *Entangled Life* is, in fact, full of descriptions of discoveries based on such chance encounters which have shaped human-fungi interactions as well as human history – for instance the discovery of penicillin and the discovery of a specific kind of penicillin mold that allowed inexpensive mass production of the drug (see Sheldrake, 208.).

79 see Haraway, "Situated Knowledges: The Science Question in Feminism and the Privilege of Partial Perspective."

80 Tsing, *The Mushroom at the End of the World*, 18.

dency to repeat cycles of destruction. Her arguments are not directed against any specific solutions, but rather against the general framing of progress narratives in its reiteration of an ingrained understanding of how humanity can and will interact with the non-human natural world. Not only is there, thus, a gradual difference between her work and that of the mycologists discussed above, but this difference emerges as a fundamental methodological one, that relies on challenging perspectives.

The chapter "Enabling Entanglements" – included at the beginning of *The Mushroom at the End of the World* – sets out an argument for the necessity of such a shift:

> Ever since the Enlightenment, Western philosophers have shown us a Nature that is grand and universal but also passive and mechanical. Nature was a backdrop and resource for the moral intentionality of Man, which could tame and master Nature. It was left to fabulists, including non-Western and non-civilizational storytellers, to remind us of the lively activities of all beings, human and not human. Several things have happened to undermine this division of labor. First, all that taming and mastering has made such a mess that it is unclear whether life on earth can continue. Second, interspecies entanglements that once seemed the stuff of fables are now materials for serious discussion among biologists and ecologists, who show how life requires the interplay of many kinds of beings. Humans cannot survive by stomping on all the others. Third, women and men from around the world have clamored to be included in the status once given to Man. Our riotous presence undermines the moral intentionality of Man's Christian masculinity, which separated Man from Nature. The time has come for new ways of telling true stories beyond civilizational first principles. Without Man and Nature, all creatures can come back to life, and men and women can express themselves without the strictures of a parochially imagined rationality. No longer relegated to whispers in the night, such stories might be simultaneously true and fabulous. How else can we account for the fact that anything is alive in the mess we have made?[81]

While I have already drawn on Tsing's argument in the preceding chapters, I have quoted it in its entirety here to highlight the connection that Tsing draws between narratives of progress, the specific relationship between "man" and "nature" they promote, and the necessity of actively engaging with the existing counternarratives to undermine the reiteration of progress narratives that continue to cement the status quo. Tsing's account does not only paint a narrative timeline of the ways in which Western cultures have understood the world within the framework of a narrative of progress since the enlightenment, she also argues that, while counternarratives

81 Tsing, vii.

have always existed at the margins of this universalizing story of "man", the fundamental separation of "man" from "nature" has led not only to misrepresentations but also to mistreatments of the non-human natural world – and to the situation which Tsing refers to as a "mess" throughout the text. This position, which she calls "humanist", in the sense that it privileges an anthropocentric perspective, pits human progress against "nature" as a romanticized space. As she argues, "[f]or humanists, assumptions of progressive mastery have encouraged a view of nature as a romantic space of anti-modernity."[82] This romanticization of "nature" emerges even in narratives of human connection or emotional engagement with "nature", as "progressive mastery" can also take the shape of attempting to make sense of or have an aesthetic reaction to the non-human natural world. Instead of striving for mastery – even the mastery of making sense of a specific environment –, Tsing "look[s] for *disturbance-based ecologies in which many species sometimes live together without either harmony or conquest*."[83]

Tsing's work described in *The Mushroom at the End of the World*, is focused on discovering and telling stories of

> precarious livelihoods and precarious environments through tracking matsutake commerce and ecology. In each case, I find myself surrounded by patchiness, that is, a mosaic of open-ended assemblages of entangled ways of life, with each further opening into a mosaic of temporal rhythms and spatial arcs. I argue that only an appreciation of current precarity as an earthwide condition allows us to notice this – the situation of our world.[84]

While this may sound like an unruly and unmanageable project, this resistance to neat structure is precisely the point of Tsing's argument, as it allows her to include different perspectives, build connections and create a network of stories without a dominant hierarchy. I will discuss Ursula Le Guin's short story "Vaster than Empires and More Slow" as a literary example of this kind of environmental narrative below. What I am interested in here is the way in which Tsing uses an anthropological project of describing human-mushroom entanglements to make sense of non-progress narratives on the "possibility of life in capitalist ruins."[85]

Human progress narratives, as Tsing argues, overshadow other world-making projects and perceptions. Her argument is based on countering such progress narratives:

---

82 Tsing, 5.

83 Tsing, 5.

84 Tsing, 4.

85 Tsing, *The Mushroom at the End of the World*.

> Progress is embedded […] in widely accepted assumptions about what it means to be human. Even when disguised through other terms, such as "agency," "consciousness," and "intention," we learn over and over that humans are different from the rest of the living world because we look forward – while other species, which live in the day to day, are thus dependent on us. As long as we imagine that humans are *made* through progress, nonhumans are stuck within this imaginative framework too.[86]

This fundamental understanding of humanity as "looking forward" underlies even Romantic engagements with "nature" that are critical of progress narratives driven by commerce and industrialization, as they are rooted in an anthropocentric perspective that privileges human agency and consciousness even where such consciousness seems to align itself with the rhythms of the non-human natural world. I have discussed several examples of Romantic perspectives critical of progress narratives and in favor of situated knowledges above and would briefly like to return to Thoreau as an example of this distinction. I have read Thoreau's criticism of the railway as a narrative of progress in *Walden* in my chapter on "Steam" above, arguing that he situates his understanding of the self in what might in Tsing's terminology be described as an entanglement with "nature", which he, however frames as a project of radical self-improvement through the interaction with "nature" and which is consequently, rooted in a deep-seated belief in human agency. Ecocritical engagements which by default speak from a human perspective must, of course, always push back against anthropocentrism if they wish to center non-human perspectives and conceptualizations of the non-human natural world, but this necessity becomes even more visible in Romantic projects that tend to *center* human autonomy and agency in the service of self-improvement or political progress. While this is not a phenomenon of American Romanticism alone, Karen Kilcup makes a similar argument for North American

> [w]omen writers in the early part of the nineteenth century [who] fought resource depletion (such as clearcutting), but they – and some male counterparts – understood that the loss was psychological, mythical, and genealogical as well as concrete. That is, "resources" were not merely extractable, salable objects, they were also crucial symbolic assets.[87]

While they urged readers towards resource conservation and argued against the practice of "clear-cutting" some of the textual examples Kilcup discusses, such as Caroline Stansbury Kirkland's in *A New Home, Who'll Follow?; or, Glimpses of Western*

86 Tsing, 21.

87 Kilcup, *Fallen Forests*, 76.

Life (1839) also struggled with the need to carry patriotic progress narratives of human improvements in and to the landscape.[88] To read these texts through an ecocritical lens, as Kilcup does, poses the additional difficulty of having to take into account "women's complicity in perspectives and agendas that complicate or even depart from today's environmental orthodoxies."[89] The complexitics of these positions may become even more pronounced if one considers that these progress narratives were, as Zilberstein has shown, often focused on the long-term project of "improving" the climate of the continent through colonization.[90]

This is not simply a question of gradual argumentative differences between historical and current ecocritical arguments, but one that draws attention to a fundamental anthropocentrism at the heart of Romantic ecological engagements which, as I have argued in my "Anthropocene" chapter, carries over into Romantic ecocriticism. Tsing's argument against "looking forward", consequently has far-reaching consequences for ecological as well as ecocritical engagements as it asks for a radical subversion of narratives of progress that goes beyond the necessity of situating knowledge in Haraway's sense, pushing at the limits of framing knowledge about the non-human natural world in literary or scientific terms. Where Haraway argues for "accountability and responsibility for translations and solidarities linking the cacophonous visions and visionary voices that characterize the knowledges of the subjugated",[91] Tsing's approach engages with the even more fundamental question of how to think about the human self and its role in the world and how to put these thoughts into language and narrative structure without recurring to progress narratives.

While I will address Ursula Le Guin's attempts at reframing narrative structures to undermine heroic and progress-oriented storytelling below, I would like to briefly turn to Tsing's argument that counter-strategies of engaging with the world may be found in how marginalized storytelling traditions frame humanity's relationship with the non-human natural world. There is a danger of taking an argumentative shortcut here – not necessarily in Tsing's argument itself, but in how it might be understood. There is a tendency in Western cultures to simply ascribe superior knowledge about "nature" to, for instance, Indigenous groups, because of an assumed "closeness" to "nature" that frames Indigenous narratives about the non-human natural world within Western binary structures. This kind of argument does not only reiterate romanticizing "noble savage" narratives, but it also transfers the necessity of knowing and caring about the natural world away from the Western

88 Kilcup, 85.

89 Kilcup, 3.

90 see Zilberstein, *A Temperate Empire*, 1–2.

91 Haraway, "Situated Knowledges: The Science Question in Feminism and the Privilege of Partial Perspective," 590.

position of "man" – positioned conveniently outside of "nature" – and onto already marginalized groups.

Part of this is, of course, a translation problem – not just between languages but also between epistemological systems. Robin Wall Kimmerer, for instance, describes her difficulty in reconciling Indigenous knowledge about plants with her scientific education in botany in *Braiding Sweetgrass* (2013) by highlighting the untranslatability of a term in Anishinaabemowin (Ojibwe) that describes an essential quality of mushrooms:

> Science can be a language of distance which reduces a being to its working parts; it is a language of objects. The language scientists speak, however precise, is based on a profound error in grammar, an omission, a grave loss in translation from the native languages of these shores. My first taste of the missing language was the word *Puhpowee* on my tongue. I stumbled upon it in a book by the Anishinaabe ethnobotanist Keewaydinoquay, in a treatise on the traditional use of fungi by our people. *Puhpowee*, she explained, translates as "the force which causes mushrooms to push up from the earth overnight." As a biologist, I was stunned that such a word existed. In all its technical vocabulary, Western science has no such term, no words to hold its mystery. You'd think that biologists, of all people, would have words for life. But in scientific language our terminology is used to define the boundaries of our knowing.[92]

Kimmerer explains that in many Native American languages, for instance the Potawatomi language of her ancestors that she is studying, there is a higher ratio of verbs (70 percent in her example), many of which may have no equivalent in English, a noun-based language, which is, as she ironically remarks, "somehow appropriate to a culture so obsessed with things".[93] While this example is particularly striking, as the word seems to describe one of the unique qualities of mushrooms that I have discussed with regard to their usage in Romantic literature above, it also draws attention to the trap of romanticizing Indigenous knowledges as more in tune with "nature" without rigorously pursuing the full implications of such an argument.

To frame this within another example: it is not a shift in perspective to repeat the common story that the Inuit have many words for snow. Instead, a more profound engagement with this knowledge would have to inquire into the purpose behind these words, the environment they describe, how this knowledge is disseminated[94] and how shifts in the environment they describe might be understood in terms of the global climate crisis which is happening at accelerated speed and thus particularly

92 Kimmerer, *Braiding Sweetgrass*, 49.

93 Kimmerer, 53.

94 see "Yes, the Inuit Have Dozens of Words for Snow – but What Does Each One Mean Exactly?"

visible in the Arctic.[95] This kind of acknowledgement of Indigenous forms of knowledge and other situated knowledge systems beyond Western structures might, indeed, contribute to a shift in perspective from universal tales of progress towards an acknowledgement of the precarity of climate crisis.[96]

In the next part of this chapter, I will take Tsing's arguments about the necessity of abandoning progress narratives to one of its roots in Ursula Le Guin's "Carrier Bag Theory of Fiction" and Le Guin's exploration of a rhizomatic perspective in her own science fiction work.

Imagining a Mushroom Perspective

Tsing distinguishes between narratives of progress, based on a temporality that is directed towards a better future and typically focusing on one heroic figure and narratives that not only allow for the inclusion of different perspectives but also might help listeners to become aware of multiple, entangled forms of experiencing the world. In this, she draws on Ursula Le Guin's "Carrier Bag Theory of Fiction", which challenges the central heroic story of hunting a mammoth to argue that "[t]he trouble is, we've all let ourselves become part of the killer story, and so we may get finished along with it. Hence it is with a certain feeling of urgency that I seek the nature, subject, words of the other story, the untold one, the life story."[97] This story, Le Guin argues, is one not based on heroic acts (killing the mammoth) but on an understanding of narrative as a container (a bag, a sack to hold grains and other gathered things) holding words or "stuff" or the things she herself brought to writing science fiction:

> [T]iny grains of things smaller than a mustard seed and intricately woven nets which then laboriously unknotted are seen to contain one blue pebble, an imperturbably functioning chronometer telling the time on another world and a mouse's skull; full of beginnings without ends, of initiations, of losses of transformations and translations, and far more tricks than conflicts, far fewer triumphs than snares and delusions, full of space ships that get stuck, missions that fail and people who don't understand.[98]

Both Tsing and Le Guin draw on feminist cultural theories that understand the world not in terms of how it can be conquered, ruled and, ultimately, destroyed, but in terms of the multitudes it may contain. The metaphor of the carrier bag, thus, also

95 see "Climate Change in the Arctic."

96 Tsing, *The Mushroom at the End of the World*.

97 Le Guin, *The Carrier Bag Theory of Fiction*, 33.

98 Le Guin, 36.

emerges as a poetics of Le Guin's own fiction, which is frequently concerned with the kind of shifts in perspective that Tsing also foregrounds in her theoretical approach to imagining the possibilities of life in capitalist ruins.

Le Guin's short story "Vaster than Empires and More Slow" (1971) is a fictionalization of this necessary shift in perspective, based on the central concepts of emotional connection and mycelial network consciousness. As in many of Le Guin's other works of science fiction, especially the texts set in the Hainish universe, it foregrounds cultural and social themes and theories such as ecocriticism and social justice.[99] One of the most interesting elements of the narrative is its multi-perspectivity and its lack of a clearly defined protagonist figure, which does not only reiterate Le Guin's thoughts on telling non-heroic stories, but also reflects the flat hierarchies and co-operative strategies of mycelial networks. This does not mean that the story is entirely without conflict, but, similar to Le Guin's ecocritical science fiction novel *The Word for World is Forest* (1972), which also shows human's attempts at colonizing and commodifying other planets, it does not present humans (and humanoid aliens) as heroic, nor do they emerge as victorious from the conflict.[100]

Instead, the crew of the spaceship exploring "World 4470" – a planet inhabited by plants but without animal life forms – is introduced, somewhat ironically, as "of unsound mind":

> Who else would go out to collect information that wouldn't be received for four, or five, or six centuries? Received by whom? [...] No sane person who has experienced time-slippage of even a few decades between near worlds would volunteer for a round trip of a half millennium. The Surveyors were escapists; misfits; nuts. Ten of them climbed aboard the ferry at Smeming Port on Pesm, and made varyingly inept attempts to get to know one another during the three days the ferry took getting to their ship, *Gum*.[101]

Part of this is, of course, an attempt at describing a distinctly unheroic group of people, in accordance with Le Guin's "Carrier Bag Theory". The repetition of "unsound mind" and "[n]o sane person", however, suggests that the narrative is explicitly creating unreliability, asking the reader not to trust these characters, who are grouped together as "[t]en of them" and labeled "inept" at building connections. The story's exposition offers no distinctions between them, nor does it privilege one perspective over another, as the narrative perspective remains heterodiegetic throughout,

99 see Barbour, "Wholeness and Balance in the Hainish Novels of Ursula K. Le Guin."

100 Incidentally, the two texts are also connected by intertextual references to Andrew Marvell's "To His Coy Mistress", and more specifically to the poem's engagement with temporality (see Marvell, "To His Coy Mistress.").

101 Le Guin, "Vaster Than Empires and More Slow," 101.

with very few instances of internal focalization, allowing rare glimpses into individual crew members' thoughts and motives.[102]

The crew of ten consists entirely of scientists, which is not only part of the irony of this construction, but also allows for a certain level of meta-reflection as their interpretation of the planet and the events is rooted in their individual scientific backgrounds. This constellation creates distance, not only between the characters, but also towards the reader, who is not invited to identify with any of the characters – human or humanoid alien – nor even with anything on the planet they explore, as this is also made strange in the narrative. This distancing strategy is further sustained by a focus on the crew's antipathy towards each other and towards one crew member in particular, the "empath" Osden.

While one of the other characters introduces Osden as having been "cured" of his "autistic condition",[103] he still possesses an unusual sense of empathy – or rather "wide-range bioemphatic receptivity" – which is the reason for his presence on board.[104] While "Mannon, the Soft Scientist (psychology, plus psychiatry, anthropology, ecology, etc.)"[105] attempts to explain Osden's condition to his fellow crew members, his explanation is framed as having gone on "for quite a long time",[106] suggesting that none of the other crew members where actually interested enough to listen, nor is the reader invited to contemplate Osden's background and develop sympathy towards him.

This climate of antipathy and misunderstanding sets the stage for the story's conflict, which pits Osden's bad behavior towards everyone else against the stinging comments of the hostile crew. Osden attempts to explain that he can't help reacting to everyone's emotions – "[m]y choice is to be hated or to be despised"[107] – but he does so in a manner that further exacerbates the conflict until he is sent away, out of empathy range, by the crew's acting "Coordinator", Tomiko Haito.[108] Distance creates further distrust, when Porlock, "the Hard Scientist (chemistry, plus physics, as-

---

102 Interestingly, the 1971 version of the story opens on the word "you", addressing the reader directly as it introduces the jarring, time- and space-rending experience of "NAFAL flight", which blurs the boundaries of the moment and the self until "[t]here is no there" and "[a]ll is here and now". The passage is no longer part of newer versions of the story and its absence underscores the sense of distance between the narrative and the reader who is no longer addressed directly (Le Guin, 99.).

103 Le Guin, 102. ("Cure" is set in quotation marks here, to signal its problematic connotations in the context of disability studies as well as the fact that the story presents it as not entirely successful and suggests, in fact, that Osden's continued empathy is a source of suffering for him, even if he is assumed to have become useful to society after having been "cured".)

104 Le Guin, 105.

105 Le Guin, 102.

106 Le Guin, 103.

107 Le Guin, 111.

108 Le Guin, 110.

tronomy, geology, etc.)",[109] reports being stalked by "something big" in the forest[110] and Osden is found unconscious and badly beaten near his camp site far away from the crew.[111] As the crew agree that the planet is a "pure phytosphere"[112] and not inhabited by animals, the fear of the unknown leads to further distrust and strife, as they struggle to make sense of the attack. Neither the "hard" nor the "soft" sciences have a useful explanation as the scientists begin to doubt their perceptions as well as their own understanding of them. Mannon, trained in psychology, reports his feelings toward the forest in a way that is at the same time scientifically accurate and useless as a solution to the problem the crew is faced with:

> I feel a strong anxiety with a specific spatial orientation. But I am no empath. Therefore, the anxiety is explicable in terms of the particular stress-situation, that is the attack on a team member in the forest, and also in terms of the total stress-situation, for which the archetypal connotations of the word 'forest' provide an inevitable metaphor.[113]

This is presented not so much a critique of the humanities, but an expression of the inability of humanist scientific and rational thought to conceptualize what lies outside of human experience and cultural knowledge, as the "Hard" sciences have already been ruled out as a source of explanation by this point. Mannon's explanation is framed as useless by the narrative itself, drawing attention to its meta-textual engagement with the question of how to shift the perspective away from anthropocentric framings of the events.

Osden, the empath, is the only person who is able to sense the fear emitted by the interconnected plant-life on the planet, but even he does not have the words to fully express what he is sensing, nor does he feel comfortable sharing his thoughts with the hostile crew. The biologist's description of the planet's interconnected phytosphere includes terms such as "root-node linkage" and suggests "incredible complexity and physical extent", but he is ultimately unable to show or explain the "connectedness" between the plants,[114] which Osden describes as "[s]entience without senses" that is "[n]othing comprehensible to an animal mind. Presence without mind. Awareness of being, without object or subject."[115]

Here, the story resonates with mycological descriptions of mycorrhizal fungal networks, even if it seems to antedate the most substantial research on the matter,

109 Le Guin, 102.
110 Le Guin, 112.
111 see Le Guin, 115.
112 Le Guin, 108.
113 Le Guin, 121.
114 Le Guin, 124.
115 Le Guin, 124.

which, according to Sheldrake, dates from the 1980s onwards. Le Guin's description may have drawn on a study conducted by Swedish botanist Erik Björkman, "who in 1960 injected trees with radioactive sugars and was able to show that radioactivity accumulated in nearby *Monotropa* plants." According to Sheldrake, this "was the first demonstration that substances might pass between plants via a fungal pathway."[116] Mycologists have termed this pathway, which exchanges information as well as nutrients and other substances, "the Wood Wide Web",[117] with some, like Stamets, arguing for fungal consciousness and comparing mycelial networks to neurological pathways.[118]

Le Guin's story, written two years after the inception of ARPA net by the U.S. Department of Defense, but still unaware of the term "World Wide Web", which was coined almost two decades later, still manages to describe the sense of disembodied (or perhaps multi-bodied) interconnectedness inherent in these metaphorical constructions. Perhaps more importantly, her story shows a shift in perspective from the "animal perspective" brought to the planet by the humanoid crew[119] to the unified plant perspective picked up by Osden, the empath: "To know the whole daylight after all, and the whole night. All the winds and the lulls together. The winter stars and the summer stars at the same time. To have roots and no enemies. To be entire. Do you see? No invasion. No others. To be whole."[120] Osden's description indicates the narrative's shift in perspective towards an ecocritical message – after all, "[t]o a forest, [...] we might appear as forest fires. Hurricanes. Dangers. What moves quickly is dangerous to a plant. The rootless would be alien, terrible."[121]

Le Guin's story – at least in its 1971 version – begins and ends in "oneness", as the time-rending experience of constantly being in the "now" during intergalactic NAFTA flight is reflected in the "oneness" of the planet's phytosphere. Osden, the empath, ultimately decides to stay on the planet, to take the fear of the other into himself and reestablish "oneness",[122] as Tomiko Haito, the coordinator, reflects in one of the few instances of clearly focalized narration: "He had given up his self to the alien, an unreserved surrender, that left no place for evil. He had learned the love of the Other, and thereby had been given his whole self. But this is not the vocabulary of reason."[123]

116 Sheldrake, *Entangled Life*, 166.

117 Sheldrake, 165–93.

118 see Stamets, *Mycelium Running*, 2–11.

119 Le Guin, "Vaster Than Empires and More Slow," 117.

120 Le Guin, 129.

121 Le Guin, 125.

122 see Watson, "The Forest as Metaphor for Mind: 'The Word for World Is Forest' and 'Vaster Than Empires and More Slow.'"

123 Le Guin, "Vaster Than Empires and More Slow," 132–33.

Tomiko's frustration at not being able to express her experience in a rational, scientific manner, reiterates the narrative's earlier critique of humanist scientific systems, which are presented as unable to frame and understand what lies outside of their anthropocentric perspective. Ian Watson argues that Le Guin's story "uses the paranormal element from the Hainish cycle as a way of validating a forest-mind which is a verisimilar actuality rather than a metaphor",[124] but I would argue that the framing is not quite as straightforward. The narrator, after all, introduces the whole crew as neurodivergent in some way and there is also no clear suggestion that a paranormal element might be the only explanation for Osden's connection with the vegetation. Moreover, the story clearly shows the crew members' frustration of not being able to make sense of their reactions to the planet. Meaning, however, still emerges from their combined perspectives.

Instead of framing Osden's integration into the planetary network as a paranormal connection, I would, therefore, like to argue that the story models a shift in perspective from the human to the non-human in its narrative construction and tone. The short passage focusing on Tomiko's attempt at understanding what happened to Osden is followed by another shift in perspective, as the heterodiegetic narrative voice resumes its description of the crew's continued presence on the planet. Something has changed, however, in the tone of the narrative, which has become more poetic and less spiteful after Osden's integration into the network, reflecting a sense of restored "oneness":

> The people of the Survey team walked under the trees, through the vast colonies of life, surrounded by a dreaming silence, a brooding calm that was half-aware of them and wholly indifferent to them. There were no hours. Distance was no matter. Had we but world enough and time ...[125]

The story circles back to its intertextual title, as it quotes the beginning of Andrew Marvell's metaphysical poem "To His Coy Mistress" (1681), but also creates an elaborate pun based on the words that are specifically not included in the title. What grows "vaster than empires and more slow" in Marvell's poem is, after all, "my vegetable love".[126] It is no coincidence that Tomiko's thoughts also circle around the concept of love, even if she rejects the idea of Osden's integration into the planetary plant organism as a form of love as ultimately irrational.

Where Marvell's poem is, however, focused on urging the "coy mistress" into giving in to the speaker's suit by arguing that they do, in fact, not have "world enough

124 Watson, "The Forest as Metaphor for Mind: 'The Word for World Is Forest' and 'Vaster Than Empires and More Slow,'" 232.

125 Le Guin, "Vaster Than Empires and More Slow," 133.

126 Marvell, "To His Coy Mistress," 1421.

and time",[127] Le Guin's story dissolves the concept of time altogether, not only in the imagination of time travel spanning several centuries, but, more importantly for my argument here, in the imagination of a oneness of connected plant life and consciousness that experiences time as a kind of instantaneous simultaneity transcending linear time. Following the intertextual reference, the narrative voice shifts from descriptions of the Survey team, who will eventually return to linear time, towards a description of the planet that seeks to verbalize a non-human experience: "The planet turned between the sunlight and the great dark; winds of winter and summer blew fine, pale pollen across the quiet seas."[128] This short passage presents the story's final shift in perspective, as it attempts to envision a narrative of existence beyond the human.

"Vaster than Empires and More Slow", thus, presents a narrative that is not driven by progress or oriented towards the future but centers on multi-perspectivity and connectedness as its driving factors, aligning itself both with Le Guin's own "Carrier Bag Theory of Fiction" and Tsing's theorization of a counter-approach to Western progress narratives. As I have argued above, these elements are also negotiated via the story's formal construction and tone, as it intentionally shows the limitations of anthropocentric perspectives – whether personal or scientific – and the failed attempts of the human characters to express consciousness beyond their own experience. I would disagree with Watson's early reading of the story that suggests that its message is one of silence, which may show a "mystic area of experience", but also shows that "there is nothing profound one can say about it."[129] Watson's perspective is focused on making rational sense of a story that is, first and foremost, an exploration of the limits of rational perspective and, thus, focused on challenging rational, humanist positionality on several levels.

A sense of narrative positionality emerges most prominently in the first paragraphs of the story – as the narrator is "trying to describe the sensations of the average person in NAFAL flight",[130] which is, perhaps, the reason why Le Guin did not include these passages (in which an overt narrative voice speaks directly to an overt addressee) in later versions of the story.[131] This attempt at dodging the positionality of narrative voice can, perhaps best be explained with Le Guin's "Carrier Bag Theory of Fiction" in mind, where she argues that novels (and potentially other types of narrative fiction such as short stories or novellas) work like containers, which "have

127 Marvell, 1420.

128 Le Guin, "Vaster Than Empires and More Slow," 133.

129 Watson, "The Forest as Metaphor for Mind: 'The Word for World Is Forest' and 'Vaster Than Empires and More Slow,'" 236.

130 Le Guin, "Vaster Than Empires and More Slow," 100.

131 see Harris, "What Makes a Great SF Short Story?" James Harris comments on this discrepancy but does not offer a conclusive explanation.

people in them" and do not focus on heroic conflict, "since [their] purpose is neither resolution nor stasis but continuing process."[132] To Le Guin, this is a feminist issue, as heroic stories of "civilization" based on the use of pointy, sharp and blunt objects as weaponry tend to exclude women or categorize them as "defective" humans or "not human at all".[133] Le Guin's project in the "Carrier Bag Theory of Fiction" as well as in her own fiction focuses on the possibilities of rethinking this story of civilization to include other ways of telling stories and other perspectives:

> If it is a human thing to do to put something you want because it's useful, edible or beautiful, into a bag, or a basket, or a bit of rolled bark or leaf, or a net woven of your own hair, or what have you, and then take it home with you, home being another, larger kind of pouch or bag, a container for people, and then later on you take it out and eat it or share it or store it up for winter in a solider container or put it in the medicine bundle or the shrine or the museum, the holy place, the area that contains what is sacred, and then next day you probably do much the same again – if to do that is human, that's what it takes, then I am a human being after all.[134]

This passage works with the same means of narrating it theorizes – highlighting and repeating various forms of containers, but also offering different forms of contextualization for these containers that do not universalize but build connections between these individual or specific cultural experiences.

Le Guin's theorization of both human civilization and the kind of stories that are usually not told about it contains a component of situated knowledge and positionality, as it draws attention to the importance of the quotidian, the repetitive, the things that are done over and over again instead of only once and with a sense of finality in mind. As I have argued in the "Anthropocene" chapter, these ways of understanding the world from the perspective of the quotidian are often dismissed as "domestic" and, therefore, less interesting than the ones that take the hero on a unique adventure in the outside world. By shifting the focus towards this kind of narrative, Le Guin also establishes a feminist positionality based on collecting and connecting that resonates through Tsing's work in *The Mushroom at the End of the World*.

In the concluding part of this chapter, I will take these ideas of rethinking narrative into another popular mushroom text – Aliya Whiteley's *The Secret Life of Fungi* – to argue that Whiteley's essayistic and autobiographical engagement with fungi differs in fundamental ways from the romanticizing narratives about fungi savior figures that I have explored above.

---

132 Le Guin, *The Carrier Bag Theory of Fiction*, 35.

133 Le Guin, 31–32.

134 Le Guin, 32–33.

In Conclusion: Mushrooms, Gender and Positionality

Mushrooms growing on uprooted tree trunk in a graveyard, Berlin, October 2021

Heiner von der Lippe, Berlin

The Prologue of Tsing's *The Mushroom at the End of the World* opens with the following question: "What do you do when your world starts to fall apart?"[135] I started this chapter in the fall of 2021, the second year of being largely stuck in place as a result of the ongoing Covid-19 pandemic. Without the ability to go anywhere in particular, I found myself following Tsing's example and going on walks – mostly in the largely unkempt areas of the local municipal graveyard – and finding mushrooms, as tenuous connections between the individual parts of this project also slowly began to take shape in my head.

Aliya Whiteley's account of fungi also begins with walks that she took with her father, showing how she gradually developed an understanding of mushrooms that is based on both exploration and connection. *The Secret Life of Fungi* is a collection of poetic essays, mixing autobiographical glimpses and mushroom facts to present an entangled narrative of her encounters with mushrooms.[136] Whiteley is an author of weird fiction, and her attempts at describing the secrets of mushrooms are often

135 Tsing, *The Mushroom at the End of the World*, 3.

136 see Whiteley, *The Secret Life of Fungi*, ix.

tinged with a strangeness that reflects not only the subject but also her own fascination with it. Individual experiences are framed by a plethora of mushroom facts, some of which appear to be triggered by chance encounters, but the central theme of the short, essayistic glimpses into Whiteley's life and the life of mushrooms is really one of making connections.

In the previous chapters, I have attempted to bring the discussion of the ways in which Romantic writing imagined specific aspects of climate change and climate catastrophe back to questions of gender and positionality and I would like to do the same here, with a look at the ways in which positionality emerges in Whiteley's text. *The Secret Life of Fungi* shows that it is possible to speak about larger and smaller connections from a personal perspective, and the ways in which mushrooms provide her with an opportunity of making such connections – or, to use an image from several of the mushroom texts I have discussed throughout – the ways in which connections seem to pop up and suggest themselves whenever she engages with mushrooms:

> [W]e live in a world of connections we don't quite make. The roads cover the continents and plane trails criss-cross the sky. The internet creates incredible opportunities for communication, for togetherness, and yet there is such loneliness in the 7 billion of us. But we are intimately linked, part of this one world, and fungi is one of the strongest glues that bind us.[137]

Whiteley's discussion continues with the familiar argument that mushrooms are, literally, everywhere, which is frequently used in mushroom documentaries and also by Stamets and Sheldrake to make audiences and readers aware of the ubiquity of fungi. However, Whiteley's use of the "glue" image resonates more with the somewhat gross stickiness of some types of mushrooms (Ophiocordyceps) that Sheldrake writes about.[138] It is an odd image that clashes with the familiarity and homeliness of the rest of the introduction, positioning Whiteley as someone who is interested in but does not romanticize fungi. The comparison of the "mushroom [as] a corporeal object" to the "spongy quality of the eye under the knife" and the "leaking fluids" as the teacher sliced it up that Whiteley draws in the first chapter supports this view.[139] For Whiteley, "[f]ungi are not like us – they are entirely, magically, something else" but at the same time fungi provide us with "insights" into a secret world: "the eruption, growth and decay overhead, inside us and under our feet."[140]

Positionality emerges in different ways in Whiteley's text. One example is her use of the pronoun "we" that shifts from describing a family unit (Whiteley and her

137 Whiteley, xi.

138 see Sheldrake, *Entangled Life*, 108.

139 Whiteley, *The Secret Life of Fungi*, 4.

140 Whiteley, xii.

father) in the Introduction to an "us all" including the reader at the end of chapter 1 and then into a scientific "we", describing the properties (and the history of categorizing) mushrooms in chapter 2. There are several other, shifting forms of "we" throughout the text, but I want to draw attention to the function of these three prominent versions that occur within the first few pages of the book here. The first one establishes Whiteley's positionality in the text and her relationship to mushrooms. The second one invites the reader to stop reading: "Let's all go on a long walk and replace words with experience. Let's go now."[141] The third one occurs after a shift from "I" – the narrative voice, who "could start the business of describing a mushroom", but then decides that "I need professional help – the services of a taxonomist" to do that.[142] After the introduction of this taxonomist (Linnaeus) the third "we" adopts the voice of a teacher, asking questions such as "[w]hat types of organisms might we expect to find in the field of fungi?" This "we" emerges not as inclusive, but as slightly pompous, in answering its own questions: "Mushrooms, of course."[143] Whiteley's text, thus, takes up different positionalities in the span of only a few pages, showing the reader different ways of engaging with fungi, as she narratively models remembering and experiencing or learning about them.

Another way in which positionality emerges in Whiteley's text beyond the autobiographical framing and the use of pronouns is the way the text personally addresses and engages readers by sharing recipes and suggestions. There is a recipe for "Slow Mushroom Stew with Cheese Dumplings" in chapter 3 of the first part "Erupt" and a detailed description of how to care for a sourdough starter in chapter 7 of part 2 "Spread", which both address the reader directly and engage with potential objections ("I promise you, it's delicious.").[144] While it would be too easy to read the sharing of recipes within the context of gendered assumptions about domesticity, there is a certain repetitiveness and familiarity in this form of address and use of language that moves far away from scientific discourse and into the realm of care that seems to resonate more with Le Guin's feminist "Carrier Bag Theory" than with its "pointy-stick" alternative.[145] The latter makes an appearance in chapter 2 of part 3 "Decay" in which a little girl, whom Whiteley describes in the third person, learns about deadly mushrooms while walking the Black Forest armed with a pointy stick. The gender dynamic of the chapter pits the girl, who "is young and small for her age" against the figure of her guide: "She is afraid of him and a little in awe: here is a man who could survive using only those hands and his knowledge."[146] As Le Guin, White-

141 Whiteley, 8.

142 Whiteley, 10.

143 Whiteley, 12.

144 Whiteley, 20. Whiteley, 95–98.

145 Le Guin, *The Carrier Bag Theory of Fiction*, 29–30.

146 Whiteley, *The Secret Life of Fungi*, 129–30.

ley identifies the ability to kill with the image of the "very large stick", suggesting that poison functions in a similar manner:[147]

> At the end of this story, the little girl develops a healthy fear of innocuous-looking things, from snakes to spiders, neighbours to nature. And she starts to write. She writes strange, not quite true, stories about softly spoken organisms with her pen, her very own little stick, clutched tight between her fingers at all times.[148]

This description of writing with pointy objects resonates with feminist literary theory, most prominently Hélène Cixous' work on the "phallogocentrism" of writing,[149] but it also suggests an attempt at writing herself into a discourse that is slightly alienating because of its gender associations. Being afraid of "neighbours" certainly sticks out as a gendered warning tale. With "her very own little stick, clutched tight between her fingers", the girl feels both inadequate and empowered – mirroring a complicated subject position very familiar to women writers, but also writers from other marginalized groups.

The Secret Life of Fungi is full of meta-narrative reflections of the shape fungal knowledge might take and how to present it. One example of this is Whiteley's discussion of her own framing of the ghost orchid, which "gets *all* of its energy from fungi [...] No photosynthesis here. It relies utterly on its underground friends: the unseen, unsung fungi." This description is immediately relativized by a narrative comment: "Forgive me if this is a romantic view." Whiteley is aware of the fact that her framing "anthropomorphise[s] and romanticise[s]" a relationship which may not be fully understood yet, relativizing her own romanticization in the process.[150] This level or reflection draws attention to a common tendency of reading fungi from an anthropocentric perspective that I have also discussed above with regard to Stamets' and Sheldrake's texts, but it does so in a self-critical manner that allows Whiteley to maintain her view "that a fungus can make a really good friend",[151] while being aware of the fact that this is a romanticization and anthropomorphization of the fungus. This level of reflection, does, however, not emerge as solipsistic or condescending but manages to broaden the discussion to include the ways in which cultural knowledge is shaped by often unacknowledged amounts of scientific work: "How easy it might seem in retrospect, looking back at the path of knowledge found, painstakingly picked out along the way."[152]

---

147 Whiteley, 134.

148 Whiteley, 135.

149 see Derrida, *Dissemination*; see Cixous, "The Laugh of Medusa."

150 Whiteley, *The Secret Life of Fungi*, 48.

151 Whiteley, 49.

152 Whiteley, 53.

The Secret Life of Fungi has a tendency to question perspectives by looking closely at connections. This occurs several times throughout the text, but perhaps most prominently in the chapter "Walking the Floor" (chapter 4 of the second part, "Spread"). Here Whiteley argues that

> [t]he peace you feel when you walk the forest floor is a lie. [...] The human perspective of a forest starts with trees. Trees: the lungs of the planet, the serene organisms living to a different, slower definition of time. What else? Birds. Birdsong clear above the shush of the leaves. And maybe some small, rustling creatures, keeping out of your way. [...] You can breathe deeply, and walk far within this illusion, while the real work of the forest goes on.[153]

While Whiteley's central concern here is to draw attention to the "[c]onstant acts of negotiation"[154] taking place in the forest and the fungal involvement in all of these processes, the text also performs a shift in perspective that does not only include the reader (the "you" walking the forest floor in the first sentence) but also a much broader "human perspective", only to return to the question of representation and Whiteley's dissatisfaction with her own and other's metaphors: "I'm not certain any of these descriptions – the trading floor, the family gathering, or even the woodwide web – accurately explains what's happening within a forest."[155] Instead, she suggests an attempt at understanding the connections through an act of making strange of the familiar:

> It might make more sense to compare it to the autonomic reactions of our own bodies. For instance, you could explain the act of taking a walk in the forest by saying: *My eyes wanted to see the trees, so they instructed my heart to send extra blood flow to my legs which traded energy, in the form of movement, in return.*[156]

By drawing attention to the oddness of this way of describing one's own human body and the interconnected organs that allow it to move and take a walk, Whiteley foregrounds the difficulty of framing a complex system such as the connection of trees and mycelia in the forest within a simple, anthropocentric perspective. There is, however, no frustration in this necessary change in perspective. As Whiteley suggests, this "is good. New words will have to be found,"[157] echoing Tsing's argument for new narratives.

153 Whiteley, 81–82.
154 Whiteley, 82.
155 Whiteley, 83.
156 Whiteley, 83.
157 Whiteley, 84.

There are several other ways in which the text addresses the ways in which stories about connections might be told – for instance the different possibilities of looking at the relationship between the ambrosia beetle and the fungus it carries, that Whiteley retells as a fairy tale from different perspectives,[158] or the reading of mushroom horror texts in chapter 6 of part 3 "Decay". One of the most prominent examples of this type of meta-textual reflection occurs in chapter 6 of part 1, which is titled "Saviours". Whiteley writes about an app on her phone "for positive news" which gives her "daily reports of progress, of happiness",[159] and describes her own reaction at such news: "On bad days, I shake my head at that two-paragraph tale of a far fetched fungus that could eat all pollution or cure a terrible disease."[160] Perspective is the central factor here, as Whiteley's reaction to these "good news" depends on her predisposition but also on the way they are framed as good news. It is, in any case, highly unlikely that a "two-paragraph tale" on an app dedicated to making people feel better by delivering "good news" would change anyone's perspective. I have discussed this narrative framing with regard to Stamets' heroic mushroom tales above, suggesting that its reframing of the fungus as an heroic figure functions as an escape mechanism for the app-user (in Whiteley's scenario), who can shift the task of saving the world towards the fungus and away from themselves. Whiteley's good-news app draws on Stamets' "big ideas" in providing readers with a sense that everything is now going to be fine, as the mushrooms will save the world.[161]

While in Tsing's work, to look for mushrooms emerges as one possible approach to a change in perspective, Whiteley gives a more literal example of the ways in which mushrooms shift perspectives in chapter 4 of part 3: "[R]esearchers at Imperial College London found that the use of psilocybin increases the perception of the user as being part of nature, one planet, at a time when many are suffering from sorrow for the world that is being lost – 'climate grief'."[162] This is, of course, not the type of solution that Tsing envisions, but it is interesting that the basis of this change in perspective is a sense of connection – even if it is one brought about by hallucinogenic mushrooms.

As may have become clear by now, this is not a chapter about mushrooms. There are, in fact very few mushrooms in it. What I have tried to do instead is to highlight some of the discourses around mushrooms that connect the Romantics' en-

158 see Whiteley, 98–104.

159 Whiteley, 33.

160 Whiteley, 33–34.

161 This is not to say that hope is useless and that we should all embrace the end of the world in a spirit of doom, but rather that the kind of bite-sized feel-good stories Whiteley describes are often counterproductive to hope-based climate activism (see, for instance, Rebecca Solnit's work, Solnit, "Why Climate Despair Is a Luxury.") in that they lead to compartmentalization and inertia.

162 Whiteley, *The Secret Life of Fungi*, 146.

gagements with them and more contemporary romanticizations to show some of the ways in which they may help to address the pressing questions of climate change and climate catastrophe. As I have argued, discourses around mushrooms in Romantic literature either frame them as plants to which other things can be compared or as mysterious entities which defy categorization and defer meaning. This latter understanding of mushrooms resonates with a more recent theorization of mushrooms as an invitation to a shift in perspective, for instance in Tsing's work, drawing on feminist figurations of situated knowledges. I have compared Ting's approach to the ways in which mushroom narratives in popular culture continue to frame mushrooms as part of heroic narratives of progress which draw on an understanding of "man"'s position in "nature" that allows for this kind of romanticization. In the last parts of this chapter, I have also introduced literary experiments at shifting perspective in Ursula Le Guin's work and brought the question of positionality into play, to show how mushroom discourses are often personal and how gender emerges as one decisive factor of engagement in these discourses. This is, however, not simply a question of perspective. As I will argue in more detail in my conclusion, the ways in which people relate to narratives about climate change and climate catastrophe directly impact their willingness to engage in mitigating behaviors. Personal factors such as class, gender, disability or geolocation do, of course, play a role, but, as I have argued with regard to mushroom-savior-narratives above, narrative patterns and framings may have a direct impact on what possibilities and necessities people are able to envision.

Conclusion

> It matters what matters we use to think other matters with; it matters what stories we tell to tell other stories with; it matters what knots knot knots, what thoughts think thoughts, what descriptions describe descriptions, what ties tie ties. It matters what stories make worlds, what worlds make stories.[1]

In the preceding chapters I have highlighted some of the discursive connections between the Romantic period and the anthropocene to draw attention to the ways in which critical conceptualizations of the early anthropocene and the current moment tend to influence each other and, more importantly, create feedback loops which reiterate dominant narratives. I have presented readings of early anthropocene texts reacting to processes of acceleration driving climate change and the impact of climate catastrophe to highlight some of the ways in which these questions emerge in Romantic poetics beyond the canon and potential ways in which they might contribute to a critical reframing of Romantic narratives that move beyond the central figures of "man" and "nature". In this concluding part of my project, I would like to bring the focus further into the present, to highlight some of the ways in which these Romanthropocene connections impact the current moment.

Two not necessarily always distinctly gendered forms of positionality emerge in my readings of early anthropocene texts in the previous chapters: one that privileges a universal or universalized position from which "the human" and "nature" are seen as distinct, often gendered entities connected by the superior understanding of a perceiving/poetic subject which situates itself as a distanced observer in order to be able to aestheticize the perceived object; and one that creates connections to human and non-human elements in "nature" to situate itself as a perceiving, often embodied poetic subject within a non-human natural world. For this second position, poetics and aesthetics are frequently rooted in situated knowledge, whereas the first

1 Haraway, *Staying with the Trouble*, 12. With thanks to Russell West-Pavlov for reminding me of this passage.

one is largely based on the assumption that the experiences of specific individuals – often critically framed as "poetic genius" or "the autonomous self" – have the power of creating and inhabiting a universalized perspective.

As I have argued in the "Apocalypse" chapter, these perspectives are gendered based on their recourse to eighteenth-century aesthetics – predominantly the Burkean distinction between the masculinized sublime subject and the feminized beautiful object. These distinctions have, to an extent, been carried over from canonized Romantic aesthetics into Romantic criticism in its attempts at synthesizing commonalities in processes of historicization and canon formation, but also in early ecocritical engagements with Romantic ecopoetics drawing on Romantic aesthetics. As I argue, these continuities manifest themselves in Romanthropocene connections and feedback loops between Romantic writing, Romantic criticism and current anthropocene discourses that privilege the narrative of the universal subject and its mastery of an externalized "nature" over situated positionalities and, thus, continue to emerge as romanticizations in the current context of climate crisis.

As I have shown in my readings of early anthropocene texts the distinction between these two emergent perspectives becomes particularly visible in Romantic texts that deal with various cultural shifts, technological acceleration processes and catastrophic events in the early anthropocene, precisely because the confrontation with these changes triggered reflection processes around humanity's relationship with "nature" in an exceedingly secularized universe. Reflecting progressive technological changes and the aftermath of catastrophic climate events on an unprecedented scale, these texts grapple with the poetic and aesthetic possibilities of expressing emotional, political or aesthetic reactions and the possibility of creating a poetic or narrative subject position to locate them in. Access to aesthetic experiences – and the connected question of whose experiences will be taken seriously and considered "valid" literary engagements with "nature" – emerge as intersecting factors, which I have attempted to address with regard to Joanna Baillie's poem "Address to a Steam-Vessel" and a brief discussion of the critical framings of Dorothy Wordsworth's work in the "Steam" chapter. These questions have gathered substantial urgency in the current moment of multiple, intersecting crises, as climate injustices affect disabled and otherwise marginalized populations on a progressively larger scale and the impact of the entangled crises of anthropogenic climate change and the ongoing Covid-19 pandemic create detrimental or even fatal health effects or function as mass-disabling events.

Early anthropocene texts draw attention to their specific cultural moment, but their attempts at negotiating the aesthetic possibilities of expressing substantial changes and catastrophes also highlight questions of positionality and universalizing narratives that continue to resonate in current anthropocene discourses. Shifts and renegotiation processes can be traced in the work of canonized as well as marginalized writers, as I have shown for Byron's "Darkness" which, in its renegotiation of the Burkean sublime, relocates the speaking subject in an imaginative dream framework; and Mary Ann Browne's "A World Without Water", which instead anchors the speaker's subject position in spatiotemporally specific embodiment. To an extent, the negotiations of a universalized subject position in Wordsworth's protest sonnet "On the Projected Kendal and Windermere Railway" can also be read as a contribution to this negotiation of positionality – albeit one that reiterates the canonized Romantics' understanding of "man"'s relationship with "nature". While these approaches are frequently gendered, based on their engagement with distinctions of the sublime and the beautiful in eighteenth-century aesthetic theory, I have also shown a counter example of the use of distinctly situated knowledge by a male-identified, canonized writer in Thoreau's engagement with the railway in *Walden*.

It would be an oversimplification to regard universalizing positions as predominantly conservative or retrospective – even if there are examples of this, such as Wordsworth's resistance to the railway – as these narratives had a far-reaching impact on Romantic ecocriticism and continue to resonate with more recent engagements with "nature" in the anthropocene. Instead, I have argued in the "Anthropocene" chapter that Romanthropocene connections between the works of the canonized Romantics, Romantic criticism and ecocriticism involved in canonization processes and historicizing anthropocene discourses that draw on early anthropocene literature and culture have created feedback loops leading to a privileging of the narrative of the universal subject and its mastery of an externalized "nature" over more diverse, situated positionalities.

As I have shown in the "Mushroom" chapter, these narratives continue to emerge as romanticizations of a specific understanding of "nature" – and humanity's role in it – in current narratives of the anthropocene; even in contexts that seem to focus on or celebrate the powers of "nature" itself. While a shift to an entirely non-human perspective can hardly be mediated in the form of human thought and writing, I have highlighted ways in which resistance to an anthropocentric perspective emerges as an anti-heroic poetics in Ursula Le Guin's "Carrier Bag Theory of Fiction" and her rhizomatic short story "Vaster than Empires and More Slow".[2] By comparison, narratives that present elements of "nature" (i.e. fungi, in the examples I discuss) as savior figures or universal remedies for the pressing problems of climate crisis often privilege an anthropocentric perspective and, thus, show similarities to narratives that rely on human inventiveness and technosolutionism[3] to "save the planet" without

2 I read Haraway's description of storytelling, quoted at the beginning of this conclusion, as a poetics of such entanglements beyond the human (see Haraway, 12.).

3 "Technosolutionism" is a portmanteau term that fuses Evgeny Morozov's conceptualization of "Technological Solutionism" – the idea that everything can be solved by inventing a new

having to change the anthropocentric focus that got us into this "mess" in the first place.[4]

As Cara Daggett summarizes drawing on Sherilyn McGregor, current eco-political strategies are frequently gendered, in the sense that technologically oriented climate mitigation has begun to display a distinct "preference for 'the kinds of solutions that are the traditional domain of men and hegemonic masculinity', which lead to a 'downgrading of ethical concerns' like justice, health, or economic equity."[5] Grassroots movements and individual ecological efforts are often feminized and, by extension, discounted as "unmanly"[6] in a Western discursive system that relies on patriarchal structures to make sense of environmental efforts. As Dagget's conceptualization of petro-masculinity shows, it is equally important to understand gendered power structures in the context of climate change denial as they foreshadow the rise of authoritarian politics drawing on concepts of hypermasculinity. In this context, climate change denial, or rather the refusal to engage with the ongoing climate crisis does not necessarily emerge from "a failure of scientific communication or reason."[7] As Dagget argues, "masculine identity, and the patriarchal orders that it supports, is important for understanding (the lack of) political responses to climate change, especially in the global North."[8] Dagget concludes that,

> [i]f people cling so tenaciously to fossil fuels, even to the point of embarking upon authoritarianism, it is because fossil fuels also secure cultural meaning and political subjectivities. Since the new imperialism of the 19th century, fossil fuels have become the metaphorical, material, and sociotechnical basis of Western petrocultures that extend across the planet.[9]

Dagget's conceptualization of the historical dimension of petrocultures reaches back to the commercial projects of industrialization and acceleration in the early anthropocene that I have traced throughout the previous chapters. While I have not

---

technical solution for it. (see Morozov, *To Save Everything, Click Here: The Folly of Technological Solutionism.*)

4 see Tsing, *The Mushroom at the End of the World.*

5 Dagget, "Petro-Masculinity: Fossil Fuels and Authoritarian Desire," 29. I would add to this that capitalist interests play a distinct role in this shift, as the commercialization of ecological crisis mitigation efforts, such as the emphasis on individual e-mobility over public transport, which is largely driven by the car manufacturing lobby in the German context, functions as another factor in the current entanglement of technosolutionism and gendered socio-political structures.

6 see Brough et al., "Is Eco-Friendly Unmanly? The Green-Feminine Stereotype and Its Effect on Sustainable Consumption."

7 Dagget, "Petro-Masculinity: Fossil Fuels and Authoritarian Desire," 41.

8 Dagget, 28–29.

9 Dagget, 27.

focused on any enthusiastic supporters of fossil fuels, Joanna Baillie's endorsement of the democratizing possibilities of steam power in "Address to a Steam-Vessel" raises important questions around the cultural meaning of fossil-fuel discourses in the anthropocene, as well as intersecting questions of access and accessibility that need to be taken into account in the context of climate justice.

Attempts at reclaiming or reframing ecological narratives that emerge from critical posthumanist, feminist, decolonial and other critical projects often highlight the problematic perseverance of the discursive separation of "man" and "nature" inherent in these perspectives, drawing attention to its exclusions based on their narrow definition of "man" and its resistance to other framings. I have highlighted some of these perspectives critical of the "anthropos" in the "Anthropocene" and the "Mushroom" chapters, arguing that this critical "stubbornness" is rooted in an understanding of "man"'s ability to achieve mastery over "nature" (in a discursive as well as a material sense) that I have also traced back to Romanticism.

More recent theorizations of the anthropocene such as Tsing's *The Mushroom at the End of the World*, have focused on potential strategies which might make it possible to move out of the constant reiteration of "narratives of progress and ruin" that dominate Western culture's understanding of the anthropocene and its exigencies. With multiple, intersecting crises evolving at the same time, an understanding of their connections and the ways in which they aggravate each other seems more than urgent. As Andrew Liu[10] argues for the connection between the Covid-19 pandemic and climate crisis, not only are these inextricably linked to narratives of industrial capitalism and progress – these principles are the driving factors behind the crises, as "rapidity [...] was built into the very design of contemporary capitalism, a creation of 'just-in-time' and 'lean' distribution methods that have quietly grown ubiquitous in recent decades" transforming the sense of time and distance.[11] There are eerie similarities between these observations and the early anthropocene engagements with steam-powered travel I have discussed in the "Steam" chapter, especially Thoreau's understanding of the railway as a commercial and socio-political project that reshaped contemporary perceptions of time and space. These reinscriptions of temporal and cultural models can create a sense of constancy or repetition in a manner that makes these developments seem inevitable. However, as I have also attempted

10 Liu's article is part of the digital humanities project *Feral Atlas*, created by Tsing et al. The project "argues that imperial and industrial infrastructures create new ecological conditions: indeed, those we identify as making the Anthropocene." *Feral Atlas* is particularly interested in the "non-design effects of infrastructure", in the sense of climate change that "is a non-designed effect of fossil-fuel burning factories, power plants, and combustion engines". The website uses "the term *Tipper* to address [...] critical transitions, that is, radical social and ecological changes" that "shift ecologies past tipping points, changing the relationship between parts and wholes". (Tsing et al., *Feral Atlas*.).

11 Liu, "Is Covid-19 a 'Just-In-Time' Pandemic?"

to point out by drawing on Malm's points against the assumed inevitability of fossil economies in the chapter on "Steam", inevitability is frequently also an element of Romanthropocene narratives that can and should be questioned. The necessity of questioning these ingrained discourses has become even more urgent in the current overlapping crises, as they are increasingly negotiated in the form of violence and the execution of power such as in wars, anti-migration policies, systems of scarcity, security and surveillance systems and the criminalization of climate activism.

Meanings shuttle back and forth across Romanthropocene connections, creating a sense of the ways in which the engagement with one period of "multifaceted rupture"[12] – to borrow Kate Rigby's term – might provide a clearer understanding of another.[13] I have traced some of these connections in the preceding chapters, by presenting gender-focused readings of early anthropocene texts in the context of specific thematic frameworks, but am also aware that more work needs to be done, especially with regard to other marginalized writers. The readings of early anthropocene texts that I have provided in the previous chapters show that the Romantic writers I discuss were largely engaged in dealing with the unforeseen and unprecedented outcomes of acceleration processes and their socio-political impact, as they attempted to redefine their own positionality in a changing world. As I have shown, these negotiations inform poetic and narrative strategies, reflected on multiple textual levels from prosodic patterns to speaker positionality and rhetorical devices to complicated frame structures and meta-textual reflections in narrative texts.

A sense of climate change or climate crisis emerges from these historically connected observations as an entangled, frequently global phenomenon. The ways in which individuals are affected by multiple climate crises and their ability to react to them are shaped by a number of intersecting positionalities – including age, gender, racializations, geopolitical location, disability and class, as well as socio-political factors connected to these positionalities. To read anthropogenic climate change and climate catastrophe as a global threat often eradicates or at least diminishes these differences. Framing the impact of climate crises and the necessary mitigation strategies against them as individual problems, however, obfuscates complex geopolitical problems and biopolitical power structures. As I have attempted to show with regard to the early anthropocene, climate discourses are driven by a wide variety of factors, including political and financial interests, but they are also rooted in an understanding of underlying terms such as "man" and "nature", which are frequently embedded in broader historical narratives.

While climate crisis is, as Morton argues, a "hyperobject" that is impossible to grasp in and of itself, this immensity should not hamper mitigation efforts. As Bould argues in *Anthropocene Unconscious*, "in the face of such debilitating immensity, we

---

12 Rigby, *Reclaiming Romanticism: Towards an Ecopoetics of Decolonization*, 3.

13 Tsing et al., *Feral Atlas*.

cannot merely shrug and take a selfie. We cannot allow the scale of the crises we are already living through, and of those to come, to trump their urgency."[14] While I agree with Bould, I would also like to draw attention to the validity of his own argument that the term "anthropocene" implicates "humanity as a geological force, but does so indiscriminately" in this context.[15] To place the focus on the "anthropos" as the driving force behind geological changes obscures, as Bould points out, the problem of scale. But to demand that "*we*" understand this leads to other problems of in- or exclusivity that I have attempted to highlight throughout.[16]

A universalizing "we" is a common perspective in anthropocene theorizations – one that often attempts to emotionally engage the reader by signaling that "we" are all in the same boat and that "we" must therefore act together.[17] Bould's argument is, of course, originally directed at one of the most pressing problems of the anthropocene – that of inaction, or rather the inaction of specific political actors. This is, in part a problem of critical framing, as Bould shows in response to Ghosh's argument in *The Great Derangement* (2017) that climate change rarely makes an appearance in fiction "taken seriously by serious literary journals".[18] However, as I have argued in the "Anthropocene" chapter, these critical mechanisms can also work as feedback loops (a term that Ghosh also uses) that reinscribe a specific understanding of whose writing about climate crisis will be taken seriously and understood as a valid engagement with themes like "nature" or "climate" in the first place.[19]

The constant cultural reinscription of specific notions of "nature" has far-reaching consequences, as I have argued in the "Anthropocene" chapter. Not only has the reiteration of discourses around "nature" in the writing of the canonized Romantics – predominantly Wordsworth – led to a narrow understanding of "nature" as an entity outside of "man", who can ultimately master it. This notion is also, as White argues, fundamental to understanding current cultural framings of and engagements with "nature", which (in British culture, at least) is seen as a "green" (and presumably

14 Bould, *The Anthropocene Unconscious – Climate Catastrophe and Culture*, 14.

15 Bould, 13.

16 see Cielemęcka and Daigle, "Posthuman Sustainability." Cielmeçka and Daigle address this question beyond the anthropocentric framework.

17 During the 2020 Covid Lockdowns, Clare Colebrook argued against Slavoj Žižek's response to the entangled crises that "we" may all be in the same storm, but we are hardly in the same boat. The observation goes back to Dipesh Chakrabarty's 2009 phrase that in the climate crisis, there would be no 'lifeboats for the rich'". (see Colebrook, "A Remarkable Brain"; see Chakrabarty, "The Climate of History: Four Theses." see also Braidotti, "'We' May Be in This Together, but We Are Not All Human and We Are Not One and the Same.").

18 Ghosh, *The Great Derangement*, 2017.

19 see Happer, "Belief in Change: The Role of Media and Communications in Driving Action on Climate Change."

pleasant) space largely to be preserved for its picturesque value.[20] This understanding of the non-human natural world hampers species preservation efforts and green energy projects alike because of its narrow understanding of "nature" and its disregard for humanity's role in shaping the specific rural landscape of Britain. As White argues, this framing draws on a feedback loop involving Romantic literature and Romantic criticism, or, more precisely, Bate's understanding of "Wordsworth's nature poetry [which] represents imaginary landscapes that have little to do with direct observation of a specific location at a specific moment in time."[21] Preservation efforts based on such a universalized understanding of "nature" may actually do more harm than good, while at the same time this understanding of "nature" as simply "green" may overshadow deeper and more complex engagements with the efforts needed to preserve biodiversity and mitigate the impact of climate crisis on a local as well as a larger scale.

As I have attempted to show with regard to the problems of positionality in early anthropocene texts, the universalizing "we", representing both the Romantic autonomous self and a universal "anthropos" in anthropocene texts runs the risk of overshadowing marginalized perspectives and reinscribing ingrained power imbalances that continue to aggravate anthropocene problems such as climate crisis, mass extinction and major health crises.[22] Instead of calling on a universalized "we" to fix these problems, I draw on feminist environmentalist criticism, arguing that the feminist approach to "situated knowledges" in Haraway's work[23] provides a theoretical model of rooting knowledge in contexts and perspectives while continually pushing against the risk of losing the possibility of epistemological theorization beyond the individual. Situating knowledge in specific socio-political contexts and intersecting positionalities reveals conflicting demands and interests, but it also provides a basis for differentiated and localized mitigation strategies. Not only does this concern the ability to understand critical distinctions in the context of climate crisis such as the gender climate gap[24] or the suppression of climate perspectives from the Global South, whose decade-long clamoring has rarely been noticed among the more dominant calls for "us" to finally do something about climate crisis.

---

20 This understanding of the landscape as "green" draws attention to the irony of repurposing Blake's critical engagement with "dark satanic mills" in *Milton* as the hymn "Jerusalem" confirming British nationalism and localized pride in "England's green and pleasant land" (Blake, "Milton," 120. see *Jerusalem and God Save the Queen – Last Night of the Proms* 2012.).

21 White, "Untidying the Landscape: Romantic Poetics, Class and Non-Human Nature," 233.

22 see Gabrielson, "The Nature of Gender," 64.

23 Haraway, "Situated Knowledges: The Science Question in Feminism and the Privilege of Partial Perspective."

24 Klus and Schramkowski use this term in analogy to the gender pay gap to refer to the intersecting problems arising from gender inequality and climate change. (see Klus and Schramkowski, "Gender Climate Gap"; see "Explainer.")

I would prefer to end this exploration of early anthropocene climate literature and its Romanthropocene connections on a hopeful note – suggesting, perhaps, that in reading the work of early anthropocene writers on climate change and climate catastrophe alongside anthropocene theorizations of more-than-human entanglements I stumbled upon an anthropocene connection that might suggest a path out of the current crises. What emerges, however, is predominantly an argument for the necessity of situated knowledges and a diversity of perspectives – in the admittedly very feminist hope that a better understanding of the narrative might help in changing it after all.

Works Cited

Abdy, Maria. "The Railway Tunnel." In *Romantic Women Poets 1788–1848* II, edited by Andrew Ashfield. Manchester: Manchester University Press, 1998. 185–86.

Aeon. "Yes, the Inuit Have Dozens of Words for Snow – but What Does Each One Mean Exactly?" *Aeon*. https://aeon.co/videos/yes-the-inuit-have-dozens-of-words-for-snow-but-what-does-each-one-mean-exactly. Accessed December 2, 2022.

Alaimo, Stacy. *Exposed: Environmental Politics and Pleasures in Posthuman Times*. Minneapolis: University of Minnesota Press, 2016.

Albright, Richard. "'In the Mean Time, What Did Perdita?': Rhythms and Reversals in Mary Shelley's *The Last Man*." *Romanticism on the Net* 13 (1999).

Aloi, Giovanni. "Editorial." *Antennae* 58 (2022). 10–11.

Anonymous. "The Last Man." *Blackwood's Edinburgh Magazine* XIX, no. CVIII (January 1826).

"Anthropocene, n. and Adj." In *OED Online*. Oxford University Press, 2014. https://www.oed.com/view/Entry/398463. Accessed November 6, 2022.

Ashfield, Andrew, ed. *Romantic Women Poets – 1770–1838*. Manchester: Manchester University Press, 1998.

Atkin, Polly. *Recovering Dorothy*. Salford: Saraband, 2022.

Bailes, Melissa. "The Psychologization of Geological Catastrophe in Mary Shelley's *The Last Man*." *ELH: English Literary History* 82, no. 2 (2015): 671–99.

Baillie, Joanna, ed. *A Collection of Poems*. London: Longman, Hurst, Rees, Orme, and Brown, 1823.

———. "Address to a Steam-Vessel." In *British Women Poets of the Romantic Era*, edited by Paula Feldman. Baltimore: The Johns Hopkins University Press, 1997. 39–43.

Banerjee, Suparna. "Beyond Biography: Re-Reading Gender in Mary Shelley's *The Last Man*." *English Studies* 91, no. 5 (2010): 519–30.

Barbour, Douglas. "Wholeness and Balance in the Hainish Novels of Ursula K. Le Guin." *Science Fiction Studies* 1, no. 3 (1974). https://www.depauw.edu/sfs/backissues/3/barbour3art.htm. Accessed November 6, 2022.

Bate, Jonathan. "Living with the Weather." *Studies in Romanticism* 35, no. 3 (1996): 431–47.

———. "Romantic Ecology Revisited." *The Wordsworth Circle* 24, no. 3 (1993): 159–62.

Batra, Nandita. "Dominion, Empathy and Symbiosis." In *The ISLE Reader: Ecocriticism 1993–2003*, edited by Michael P. Branch and Scott Slovic. Athens: University of Georgia Press, 2003. 101–20.

Beer, John. "Romantic Apocalypses." *The Wordsworth Circle* 32, no. 2 (2001): 109–16.

Behringer, Wolfgang. "Die Tamborakrise." In *1816 – Das Jahr Ohne Sommer*, edited by Senta Herkle, Sabine Holtz, and Gert Kollmer-von Oheimb-Loup. Stuttgart: Kohlhammer, 2019. 5–47.

Berendt, Stephen C. "Poetry." In *The Cambridge Companion to Women's Writing in the Romantic Period*, edited by Devoney Looser. Cambridge: Cambridge University Press, 2015. 1–15.

Blake, William. "Milton." In *Romanticism*, edited by Duncan Wu. Oxford: Blackwell, 1998. 120.

Blamire, Susanna. "Stocklewath; or, The Cumbrian Village." In *British Women Poets of the Romantic Era*, edited by Paula Feldman. Baltimore: The Johns Hopkins University Press, 1997. 115–46.

Boström, Magnus, and Debra J. Davidson, eds. *Environment and Society*. Cham: Springer, 2018.

Bould, Mark. *The Anthropocene Unconscious – Climate Catastrophe and Culture*. London: Verso, 2021.

Braidotti, Rosi. "'We' May Be in This Together, but We Are Not All Human and We Are Not One and the Same." *Ecocene: Cappadocia Journal of the Environmental Humanities* 1, no. 1 (2020): 26–31.

Briggle, Adam. *Thinking Through Climate Change*. Cham: Palgrave Macmillan, 2021.

Brough, Aaron R., James E. B. Wilkie, Jingjing Ma, Matthew S. Isaac, and David Gal. "Is Eco-Friendly Unmanly? The Green-Feminine Stereotype and Its Effect on Sustainable Consumption." *Journal of Consumer Research* 43, no. 4 (December 2016): 567–82.

Browne, Mary Anne. "A World Without Water." In *British Women Poets of the Romantic Era*, edited by Paula Feldman. Baltimore: The Johns Hopkins University Press, 1997. 155–59.

Buell, Lawrence. *The Environmental Imagination*. Cambridge MA: The Belknap Press, 1995.

Burke, Edmund. *A Philosophical Enquiry into the Origin of Our Ideas of the Sublime and the Beautiful*. Oxford: Oxford University Press, 1990.

Byron, George Gordon, Lord. "Darkness." In *Romanticism*, edited by Duncan Wu. Oxford: Blackwell, 1998. 716–17.

Chakrabarty, Dipesh. "The Climate of History: Four Theses." *Critical Inquiry* 35, no. 2 (2009): 197–222.

Chander, Manu Samriti. *Brown Romanticism – Poetry and Nationalism in the Global Nineteenth Century*. Lewisburg: Bucknell University Press, 2017.

Chatterjee, Ranita. "Our Bodies, Our Catastrophes: Biopolitics in Mary Shelley's *The Last Man*." *European Romantic Review* 25, no. 1 (2014): 35–49.
Cielemęcka, Olga, and Christine Daigle. "Posthuman Sustainability: An Ethos for Our Anthropocenic Future." *Theory, Culture & Society* 36, no. 7–8 (2019): 67–87.
Cixous, Hélène. "The Laugh of Medusa." *Signs* 1, no. 4 (1976): 875–93.
Clemens, Justin. *The Romanticism of Contemporary Theory: Institutions, Aesthetics, Nihilism*. Aldershot, Ashgate, 2003.
Climate Change in the Arctic. https://www.npolar.no/en/themes/climate-change-in-the-arctic/. Accessed November 6, 2022.
Colebrook, Claire. "A Remarkable Brain – Lockdown Theory #29." *Identities – Journal for Politics, Gender and Culture*, 2020. https://identitiesjournal.edu.mk/index.php/IJPGC/announcement/view/39. Accessed November 6, 2022.
Coleridge, Samuel Taylor. "Letter from S.T. Coleridge to Robert Southey, 11 September 1803 (Including early version of 'The Pains of Sleep')." In *Romanticism*, edited by Duncan Wu. Oxford: Blackwell, 1998. 512–13.
Collett, Anne, and Olivia Murphy, eds. *Romantic Climates: Literature and Science in an Age of Catastrophe*. Cham: Springer, 2019.
Cove, Patricia. "'The Earth's Deep Entrails': Gothic Landscapes and Grotesque Bodies in Mary Shelley's *The Last Man*." *Gothic Studies* 15, no. 2 (2013): 19–36.
Craciun, Adriana. *Fatal Women of Romanticism*. Cambridge: Cambridge University Press, 2002.
Crutzen, Paul J., and Eugene F. Stoermer. "The 'Anthropocene.'" *IGBP Global Change Newsletter* 41 (2000): 17–18.
Curran, Stuart. "Romantic Poetry: The I Altered." In *Romantic Writings*. London: Routledge, 1996. 185–207.
Curtis, Abi. "Mushrooming: Resistance and Creativity in Sigmund Freud and Emily Dickinson." *Angelaki* 18, no. 2 (2013): 29–44.
Dagget. "Petro-Masculinity: Fossil Fuels and Authoritarian Desire." *Millennium* 47, no. 1 (2018): 25–44.
Davies, Jeremy. "Introduction: Romantic Studies and the 'Shorter Industrial Revolution.'" *Studies in Romanticism* 61 (2022): 187–202.
———. "Romantic Ecocriticism: History and Prospects." *Literature Compass* 15, no. 9 (2018): e12489.
———. *The Birth of the Anthropocene*. Oakland: University of California Press, 2016.
Davis, Heather, and Zoe Todd. "On the Importance of a Date, or Decolonizing the Anthropocene." *ACME* 16, no. 4 (2017): 761–80.
Day, Aidan. *Romanticism*. Abingdon: Routledge, 1996.
Derrida, Jacques. *Dissemination*. Chicago: Chicago University Press, 1991.
Despotopoulou, Anna. *Women and the Railway, 1850–1915*. Edinburgh: Edinburgh University Press, 2022.

Dickinson, Emily. "The Mushroom Is the Elf of Plants." In *The Complete Poems of Emily Dickinson*. Boston: Little, Brown and Company, 1960. 565–66.
Dingley, R.J. "'I Had a Dream…' Byron's 'Darkness.'" *The Byron Journal* 9 (1981): 20–33.
Engh, Catherine. "Wordsworth, Ecocriticism, and Natural Education." *European Romantic Review* 34, no. 6 (2023): 673–87.
Fantastic Fungi. Netflix, 2019.
Feldman, Paula, ed. *British Women Poets of the Romantic Era*. Baltimore: The Johns Hopkins University Press, 1997.
Ferber, Michael. *The Cambridge Introduction to Romantic Poetry*. Cambridge: Cambridge University Press, 2012.
Fiorentini, Luca. "Mary Shelley's *The Last Man* and Post-Apocalyptic Writing." *Anglistica Pisana* 13, no. 1–2 (2016): 51–63.
Ford, Thomas H. "The Romanthropocene." *Literature Compass* 15, no. 5 (2018): e12464.
Fungi Perfecti. "Host Defense® Mushrooms." https://fungi.com/collections/host-defense. Accessed December 2, 2022.
Gabrielson, Teena. "The Nature of Gender." In *The Cambridge Companion to Environmental Humanities*, edited by Jeffrey Jerome Cohen and Stephanie Foote. Cambridge: Cambridge University Press, 2021. 56–69.
Gan, Elaine, Anna Lowenhaupt Tsing, Heather Swanson, and Nils Bubandt. "Haunted Landscapes of the Anthropocene." In *Arts of Living on a Damaged Planet*, edited by Elaine Gan, Anna Lowenhaupt Tsing, Heather Swanson and Nils Bubandt. Minneapolis: University of Minnesota Press, 2017. G1–14.
"Genius, n. and Adj." In *OED Online*. Oxford University Press. https://www.oed.com/view/Entry/77607. Accessed November 6, 2022.
Ghosh, Amitav. *The Great Derangement*. Chicago: University of Chicago Press, 2017.
Gidal, Eric. "Industrial Transport and Political Economy in *Blackwood's Edinburgh Magazine*." *Studies in Romanticism* 61, no. 2 (2022): 279–303.
Gottlieb, Evan. "Anthropocene Temporalities and British Romantic Poetry." In *Romanticism and Time*, edited by Sophie Laniel-Musitelli and Céline Sabiron. Cambridge: Open Book Publishers, 2021. 25–47.
Grasswick, Heidi. "Climate Change Science and Responsible Trust: A Situated Approach." *Hypatia* 29, no. 3 (2014): 541–57.
Haefner, Joel. "The Romantic Scene(s) of Writing." In *Re-Visioning Romanticism*, edited by Carol Shiner Wilson and Joel Haefner. Philadelphia: University of Pennsylvania Press, 1995. 256–73.
Haekl, Ralf. *Handbook of British Romanticism*. Berlin: De Gruyter, 2017.
Happer, Catherine. "Belief in Change: The Role of Media and Communications in Driving Action on Climate Change." *Climate Change and the Humanities*, edited by Alexander Elliott, James Cullis, and Vinita Damodaran. London: Palgrave Macmillan, 2017. 177–97.

Haraway, Donna. "Anthropocene, Capitalocene, Chthulhucene." In *Art in the Anthropocene*, edited by Heather Davis and Etienne Turpin. London: Open Humanities Press, 2015. 255–70.

———. "Situated Knowledges: The Science Question in Feminism and the Privilege of Partial Perspective." *Feminist Studies* 14, no. 3 (1988): 575–99.

———. *Staying with the Trouble*. Durham and London: Duke University Press, 2016.

Harris, James W. "What Makes a Great SF Short Story?" *Classics of Science Fiction*, 19 July 2020. https://classicsofsciencefiction.com/2020/07/19/what-makes-a-great-sf-short-story/. Accessed November 6, 2022.

Hathaway, Michael J. *What a Mushroom Lives For*. Princeton: Princeton University Press, 2022.

Hay, Daisy. *Young Romantics*. London: Bloomsbury, 2010.

Heald, George. "Reply to Wordsworth." https://www.songsfromtheageofsteam.uk/note-560-1/122-railways/advent-of-the-railway/first-reactions-to-the-railway/2643-663a. Accessed September 11, 2022.

Heise-von der Lippe, Anya. *Monstrous Textualities*. Cardiff: University of Wales Press, 2021.

———. "Cripping the Neoliberal University – We Need a Politics of Care." *Trafo Blog for Transregional Research*. 10 February 2022, https://trafo.hypotheses.org/33018. Accessed January 21, 2024.

Henning, Peter. "Romantic Fungi and Other Useless Things: Arnold, Tieck, Keats." *Romanticism* 26, no. 3 (2020): 292–302.

Higgins, David. *British Romanticism, Climate Change and the Anthropocene – Writing Tambora*. Cham: Palgrave Macmillan, 2017.

———, ed. *Romantic Genius and the Literary Magazine: Biography, Celebrity, Politics*. London: Routledge, 2012.

Hird, Myra J. "Proliferation, Extinction and an Anthropocene Aesthetic." In *Posthumous Life*, edited by Jami Weinstein and Claire Colebrook. New York: Continuum, 2017. 251–69.

Hoedt, Madelon. "Gothic Drama and the Uncanny Stage." In *The Palgrave Handbook of Steam Age Gothic*, edited by Clive Bloom. Cham: Palgrave, 2021. 453–64.

"Home." *Romantic Circles*. https://romantic-circles.org/#. Accessed January 5, 2024.

Hsu, Li-Hsin. "The Romance of Transportation in Wordsworth, Emerson, De Quincey, and Dickinson." *Romanticism* 25, no. 1 (2019): 45–57.

Hume, Angela, and Gillian Osborne. *Ecopoetics: Essays in the Field*. Iowa City: University of Iowa Press, 2018.

IEP. *Ecological Threat Register*. Sydney: Institute for Economics & Peace, 2020.

Jameson, Fredric. *The Political Unconscious*. Ithaca: Cornell University Press, 1981.

Jarrett, Michael. "*Walden* + Railroad + Sound." *Rhizome* 18 (2008): 1–28.

Jay, Mike. "Fungi, Folklore, and Fairyland." *The Public Domain Review*, 2020. https://publicdomainreview.org/essay/fungi-folklore-and-fairyland/. Accessed January 5, 2024.

Jerusalem and God Save the Queen – Last Night of the Proms 2012. https://www.youtube.com/watch?v=041nXAAn714. Accessed January 5, 2024.

Johnson, Thomas. "Note on Poem 1298." *Emilydickinson.org*, 1998. http://archive.emilydickinson.org/correspondence/norcross/jnp1298.html. Accessed January 5, 2024.

Keats, John. *Endymion*. London: Taylor and Hessey, 1818.

Kent, Elizabeth. *Flora Domestica: Or, The Portable Flower-Garden*. London: Taylor and Hessey, 1823.

Kerridge, Richard. "An Ecocritic's Macbeth." In *Ecocritical Shakespeare*, edited by Lynne Dickson Bruckner and Daniel Brayton. Aldershot, Ashgate, 2011. 193–210.

Kilcup, Karen L. *Fallen Forests: Emotion, Embodiment, and Ethics in American Women's Environmental Writing, 1781–1924*. Athens: University of Georgia Press, 2013.

Kimmerer, Robin Wall. *Braiding Sweetgrass*. Minneapolis: Milkweed, 2013.

Klus, Sebastian, and Barbara Schramkowski. "Gender Climate Gap." In *Klimakrise, Sozialökologischer Kollaps und Klimagerechtigkeit*, edited by Tino Pfaff, Barbara Schramkowski, and Ronald Lutz. Weinheim: Beltz, 2022. 229–241.

Knellwolf, Christa and Jane Goodall. *Frankenstein's Science – Experimentation and Discovery in Romantic Culture, 1780–1830*. Aldershot: Ashgate, 2008.

Knowles, Claire. "Female Romantic Poetry, 1798–1819: The Climate of Fear and the Loss of a Radical Generation." *Women's Writing* 28, no. 3 (2021): 305–19.

Landon, Letitia Elizabeth. "Love's Last Lesson." In *The Golden Violet*. London: Longman, Rees, Orme, Brown, and Green, 1827. 298–306.

———. "The Factory." In *Romantic Women Poets 1770–1838*, edited by Andrew Ashfield. Manchester: Manchester University Press, 1995. 213–15.

Lappin, A. Graham. "The Loch Lomond Steamers." *The Vale of Leven*. http://www.valeofleven.org.uk/lochlomondsteamers.html. Accessed August 26, 2022.

Le Guin, Ursula K. *The Carrier Bag Theory of Fiction*. Ignota, 2019.

———. "Vaster Than Empires and More Slow." In *New Dimensions* 1, edited by Robert Silverberg. New York: Avon, 1973. 99–134.

Lee, Jae Rhim. "TED: My Mushroom Burial Suit." 2011. https://www.ted.com/talks/jae_rhim_lee_my_mushroom_burial_suit?language=en. Accessed January 5, 2024.

Lennartz, Norbert. "Probing the Realms of Lostness, Non-Canonicity and Oblivion: An Introduction." In *The Lost Romantics*, edited by Norbert Lennartz. Cham: Palgrave Macmillan, 2020. 3–18.

Liu, Andrew B. "Is Covid-19 a 'Just-In-Time' Pandemic?" In *Feral Atlas: The More-Than-Human Anthropocene*, edited by Anna Lowenhaupt Tsing, Jennifer Deger, Alder

Kelemann Saxena, and Feifei Zhou. Stanford CA: Stanford University Press, 2020. https://feralatlas.org/. Accessed November 6, 2023.

Logan, Maria. "Verses on Hearing That an Airy and Pleasant Situation, near a Populous and Commercial Town Was Surrounded with New Buildings." In *British Women Poets of the Romantic Era*, edited by Paula Feldman. Baltimore: The Johns Hopkins University Press, 1997. 442.

Lorke, Melanie. *Liminal Semiotics: Boundary Phenomena in Romanticism*. Berlin: Akademie Verlag, 2013.

Lyons, Emily. "'O Little Isle!': Landscape, Englishness, and Apocalypse in *The Last Man* and *The War of the Worlds*." *ISLE* 27, no. 3 (2020): 472–90.

Malm, Andreas. *Fossil Capital*. London: Verso, 2020.

Mangrum, Benjamin. "Nature, Necessity, and the Philosophy of Metaphor in *Walden*." *Nineteenth Century Prose* 44, no. 2 (2017): 49–70.

Mann, Bonnie. *Women's Liberation and the Sublime: Feminism, Postmodernism, Environment*. Oxford: Oxford University Press, 2006.

Marshall, Bridget M. *Industrial Gothic*. Cardiff: University of Wales Press, 2021.

Marvell, Andrew. "To His Coy Mistress." In *The Norton Anthology of English Literature* 1. New York: Norton, 1993. 1420–21.

Masson-Delmotte, Valerie, Anna Pirani, and Sarah L. Connors. "Climate Change 2021: The Physical Science Basis." *IPCC*. Cambridge: Cambridge University Press, 2021.

McCarthy, Jeffrey Mathes. "Beyond Romantic Nature: Ecocriticism's New Shades of Green." *Journal of Environmental Studies and Sciences* 2, no. 3 (2012): 278–84.

McGann, Jerome J. *The Romantic Ideology: A Critical Investigation*. Chicago: University of Chicago Press, 1985.

McRuer, Robert. *Crip Times*. New York: New York University Press, 2018.

Mee, John. "'All That the Most Romantic Imagination Could Have Previously Conceived': Writing an Industrial Revolution, 1795 to 1835." *Studies in Romanticism* 61, no. 2 (2022): 229–54.

Mehnert, Antonia. *Climate Change Fictions*. Cham: Palgrave Macmillan, 2016.

Mellor, Anne K. *Mary Shelley*. New York: Methuen, 1988.

———. *Romanticism and Gender*. New York: Routledge, 1993.

———. "Thoughts on Romanticism and Gender." *European Romantic Review* 23, no. 3 (2012): 343–48.

Mishra, Vijay. *The Gothic Sublime*. New York: SUNY Press, 1994.

Mondello, Kaitlin. "The 'Grim Unreality': Mary Shelley's Extinction Narrative in *The Last Man*." *Essays in Romanticism* 24, no. 2 (2017): 163–79.

Money, Nicholas P. *Fungi – A Very Short Introduction*. Oxford: Oxford University Press, 2016.

Moore, Bryan L. *Ecological Literature and the Critique of Anthropocentrism*. Cham: Palgrave Macmillan, 2017.

Moretti, Franco. *Distant Reading*. London: Verso, 2015.
Morozov, Evgeny. *To Save Everything, Click Here: The Folly of Technological Solutionism*. New York: PublicAffairs, 2013.
Morton, Timothy. *Hyperobjects*. Minneapolis: University of Minnesota Press, 2013.
MSNBC – Senator Throws Snowball in the Senate. MSNBC, 2015. https://youtu.be/NxU55cEamco?si=xzTFEOYE3sryiYuX. Accessed January 5, 2024.
Mulvihill, James. "Consuming Nature: Wordsworth and the Kendal and Windermere Railway Controversy." *Modern Language Quarterly* 56, no. 3 (1995): 305–27.
Murphy, Conor, Robert L. Wilby, Tom Matthews, Csaba Horvath, Arlene Crampsie, Francis Ludlow, Simon Noone, et al. "The Forgotten Drought of 1765–1768: Reconstructing and Re-Evaluating Historical Droughts in the British and Irish Isles." *International Journal of Climatology* 40 (2020): 5329–51.
Murphy, Olivia. "Apocalypse Not Quite." In *Romantic Sustainability*, edited by Ben P. Robertson. Washington: Lexington, 2015. 245–59.
Newlyn, Lucy. *William and Dorothy Wordsworth: "All in Each Other."* Oxford: Oxford University Press, 2013.
O'Connell, Anita. "Visions in Verse: Writing the Visual in Romantic Dream Visions." *Studies in the Literary Imagination* 48, no. 1 (2015): 35–54.
Page, Thomas, and Gisella Deputato. "Dying to Be Green: Are Mushroom Coffins the Secret to an Eco-Friendly Death?" *CNN*, November 17, 2021. https://www.cnn.com/2021/11/17/europe/loop-mycelium-mushroom-coffin-eco-funeral-spc-intl/index.html. Accessed January 5, 2024.
Paley, Morton D. *Apocalypse and Millennium in English Romantic Poetry*. Oxford: Clarendon, 1999.
———. "Apocalypse Without Millennium." *Romantic Circles*, reprinted from the *Keats-Shelley Review* 4 (Autumn 1989), 1–25, October 1, 1997. https://romantic-circles.org/editions/mws/lastman/paley.htm. Accessed November 6, 2022.
———. "Byron." In *Apocalypse and Millennium in English Romantic Poetry*, 1–27. Oxford: Clarendon, 1999.
Payne, Geoff. *Dark Imaginings: Ideology and Darkness in the Poetry of Lord Byron*. Bern: Peter Lang, 2008.
Perry, Imani. *Vexy Thing – On Gender and Liberation*. Durham: Duke University Press, 2018.
Phillips, James. "Metaphor and the Unprecedented: Byron's 'Darkness' and Responding to Ecological Disaster." In *Romantic Climates: Literature and Science in an Age of Catastrophe*, edited by Anne Collett and Olivia Murphy. Cham: Springer, 2019. 159–71.
Plumwood, Val. *Feminism and the Mastery of Nature*. London: Routledge, 1993.
Pozoukidis, Konstantinos. "The Survival of Non-Productive Labour in Mary Shelley's *The Last Man*." *Eighteenth-Century Fiction* 33, no. 3 (2021): 393–412.
Primack, Richard B. *Walden Warming*. Chicago: University of Chicago Press, 2014.

Pyne, James Baker. "Lake Windermere from Orrest Head." *Dove Cottage and The Wordsworth Museum*, 1849. https://artuk.org/discover/artworks/lake-windermere-from-orrest-head-143072. Accessed January 5, 2024.

Recovering Dorothy: Dorothy Wordsworth after the Grasmere Journals, 2022. https://www.youtube.com/watch?v=-WiYF1r2MIA. Accessed January 5, 2024.

Reno, Seth T. *Early Anthropocene Literature in Britain – 1750–1884*. Cham: Palgrave Macmillan, 2020.

"Retirement, n.: Oxford English Dictionary." https://oed.com/view/Entry/164330?redirectedFrom=retirement#eid. Accessed December 5, 2020.

Richardson, Rebecca. "The Environmental Uncanny: Imagining the Anthropocene in Mary Shelley's *The Last Man*." *ISLE* 26, no. 4 (2019): 1062–83.

Rigby, Kate. *Reclaiming Romanticism: Towards an Ecopoetics of Decolonization*. London: Bloomsbury Academic, 2022.

———. "Discourses of Nature." In *The Cambridge History of European Romantic Literature*, edited by Patrick Vincent. Cambridge: Cambridge University Press, 2023. 73–104.

Robinson, Mary Elizabeth. "A London Summer Morning." In *Nineteenth-Century Women Poets*, edited by Isobel Armstrong and Joseph Bristow. Oxford: Clarendon, 1996. 183–84.

Rouhette, Anne. "Gender Stereotypes in Some Romantic Travelogues (and How to Use Them)." *E-Rea. Revue Électronique d'études Sur Le Monde Anglophone* 14, no. 1 (2016): 1–14.

Ruland, Richard, and Malcolm Bradbury. *From Puritanism to Postmodernism – A History of American Literature*. New York: Penguin, 1992.

Ruppert, Timothy. "Time and the Sibyl in Mary Shelley's *The Last Man*." *Studies in the Novel* 41, no. 2 (2009): 141–56.

Ruskin, John. "The Storm-Cloud of the Nineteenth Century." In *The Norton Anthology of English Literature* 2, edited by M.H. Abrams. New York: Norton, 1993. 1290–97.

Sandy, Mark. "Defence of Poetry." *The Literary Encyclopedia*, 2004. https://www.litencyc.com/php/sworks.php?rec=true&UID=5662. Accessed January 5, 2024.

Sargent, Andrew. "Citation and the No Future of Romanticism in Mary Shelley's *The Last Man*." *European Romantic Review* 31, no. 3 (2020): 313–24.

Schor, Esther. "'Stirring Shades': The Romantic Ode and Its Afterlives." In *A Companion to Romantic Poetry*, edited by Charles Mahoney. Chichester: Wiley Blackwell, 2011. 107–22.

Schroeder, Ronald A. "Byron's 'Darkness' and the Romantic Dis-Spiriting of Nature." In *Approaches to Teaching Byron's Poetry*, edited by Frederick W. Shilstone. New York: The Modern Language Association of America, 1991. 113–18.

Scott, Heidi C. M. *Chaos and Cosmos: Literary Roots of Modern Ecology in the British Nineteenth Century*. University Park: Penn State Press, 2014.

Sheldrake, Merlin. *Entangled Life*. London: The Bodley Head, 2020.

———. "Merlin Sheldrake – Videos." *Merlin Sheldrake*. https://www.merlinsheldrake.com/videos. Accessed October 24, 2022.

———. "Oyster Mushrooms Sprouting from Merlin Sheldrake's Book, *Entangled Life*." 2020. https://www.youtube.com/watch?v=chOsO7XhLB8. Accessed January 5, 2024.

Shelley, Mary. *Frankenstein*. New York: Norton, 2012.

———. *The Last Man*. Oxford: Oxford University Press, 1994.

Shelley, Percy Bysshe. "A Defence of Poetry." In *Romanticism*, edited by Duncan Wu. Oxford: Blackwell, 1998. 944–56.

———. "Mont Blanc." In *Romanticism*, edited by Duncan Wu. Oxford: Blackwell, 1998. 845–49.

Smethurst, Paul. "Prescribing Nature: William Wordsworth's *Guide Through the Lakes*." In *Travel Writing and the Natural World, 1768–1840*, edited by Paul Smethurst. London: Palgrave Macmillan, 2012. 171–80.

Smith, Charlotte Turner. *The Emigrants*. London: T. Cadell, 1793.

Solnit, Rebecca. "The myth of 'Thoreau's mother did his laundry' will not die." Facebook, September 20, 2021. https://www.facebook.com/rebecca.solnit/posts/pfbid02R4VdPtdaPs8fsvvn75LUXU3sq6u8HkbuzJzV78zPGHUBaJz9MxRXjpGG3KCLovZAl. Accessed January 5, 2024.

———. "Why Climate Despair Is a Luxury." *New Statesman*, July 17, 2023. https://www.newstatesman.com/environment/2023/07/rebecca-solnit-climate-despair-hope. Accessed January 5, 2024.

Solomonescu, Yasmin. "Mary Shelley's Fascinations: *The Last Man*." *Modern Philology* 114, no. 3 (2017): 702–25.

"Sordid, Adj. and n." In *OED Online*. Oxford University Press. https://www.oed.com/view/Entry/184862. Accessed November 8, 2022.

Southey, Robert. "681. Robert Southey to John Rickman, 2 June 1802." *Romantic Circles*, August 1, 2011. https://romantic-circles.org/editions/southey_letters/Part_Two/HTML/letterEEd.26.681.html. Accessed November 6, 2022.

———. "The Collected Letters of Robert Southey." *Romantic Circles*, 2009. https://romantic-circles.org/editions/southey_letters. Accessed November 6, 2022.

———. "To Margaret Hill." In *The Poetical Works of Robert Southey*. London: Longman, Brown, Greene and Longmans, 1844. 139–45.

Stabler, Jane. "1816 in the Manuscripts of Lord Byron and Mary Shelley." *Keats-Shelley Journal* 66 (2017): 55–76.

Stamets, Paul. *Mycelium Running*. Berkeley: Ten Speed Press, 2005.

———. "TED: 6 Ways Mushrooms Can Save the World." 2008. https://www.ted.com/talks/paul_stamets_6_ways_mushrooms_can_save_the_world. Accessed January 5, 2024.

Star Trek Discovery: "The Butcher's Knife Cares Not for the Lambs Cry." CBS, 2017.

Stevens, Lara, Peta Tait, and Denise Varney, eds. *Feminist Ecologies*. Cham: Springer, 2018.

Stommel, Henry, and Elizabeth Stommel. *Volcano Weather*. Newport: Seven Seas Press, 1983.

Tee, Ve-Yin, ed. *Romantic Environmental Sensibility*. Edinburgh: Edinburgh University Press, 2022.

The Day After Tomorrow, 20th Century Fox, 2004.

Thoreau, Henry David. "Journal – June 23." In *The Journals of Henry David Thoreau*, edited by Bradford Torrey and Francis H. Allen, V:295–99. Boston: Houghton Mifflin, 1949.

———. *The Succession of Forest Trees and Wild Apples*. Boston: Houghton Mifflin, 1863.

———. *Walden*. New York: Penguin, 1986.

Tsing, Anna Lowenhaupt. *The Mushroom at the End of the World*. Princeton: Princeton University Press, 2015.

Tsing, Anna Lowenhaupt, Jennifer Deger, Alder Kelemann Saxena, and Feifei Zhou, eds. *Feral Atlas: The More-Than-Human Anthropocene*. Stanford: Stanford University Press, 2020. https://feralatlas.org/. Accessed November 6, 2023.

UN Women – Headquarters. "Explainer: How Gender Inequality and Climate Change Are Interconnected." https://www.unwomen.org/en/news-stories/explainer/2022/02/explainer-how-gender-inequality-and-climate-change-are-interconnected. Accessed December 3, 2022.

Vail, Jeffrey. "'The Bright Sun Was Extinguish'd': The Bologna Prophecy and Byron's 'Darkness.'" *The Wordsworth Circle* 28, no. 3 (1997): 183–92.

Vargo, Lisa. "Anna Barbauld and Natural Rights: The Case of 'Inscription for an Ice-House.'" *European Romantic Review* 27, no. 3 (2016): 331–39.

Wang, Fuson. "We Must Live Elsewhere: The Social Construction of Natural Immunity in Mary Shelley's *The Last Man*." *European Romantic Review* 22, no. 2 (2011): 235–55.

Watson, Ian. "The Forest as Metaphor for Mind: 'The Word for World Is Forest' and 'Vaster Than Empires and More Slow.'" *Science Fiction Studies* 2 (1975): 231–37.

Wells, John Edwin. "Wordsworth and Railways in 1844–1845." *Modern Language Quarterly* 6, no. 1 (March 1, 1945): 35–50.

West-Pavlov, Russell. *Temporalities*. London: Routledge, 2013.

White, Simon J. "Untidying the Landscape: Romantic Poetics, Class and Non-Human Nature." In *Romantic Environmental Sensibility*, edited by Ve-Yin Tee. Edinburgh: Edinburgh University Press, 2022. 230–48.

Whiteley, Aliya. *The Secret Life of Fungi*. London: Elliott and Thompson, 2020.

Whitney, Julian S. "Planetary Crisis: Consumption and Resource Management in Byron's 'Darkness.'" *The Byron Journal* 50, no. 1 (2022): 35–48.

Wilson, Carol Shiner, and Joel Haefner. "Introduction." In *Re-Visioning Romanticism*, edited by Carol Shiner Wilson and Joel Haefner. Philadelphia: University of Pennsylvania Press, 1995. 1–14.

Wollstonecraft, Mary. "On Poetry." In *Women's Writing of the Romantic Period 1789–1836: An Anthology*, edited by Harriet Devine Jump. Edinburgh: Edinburgh University Press, 1997. 48–52.

Wood, Gillen D'Arcy. *Tambora*. Princeton: Princeton University Press, 2014.

Woolf, Virginia. *A Room of One's Own*. London: Penguin, 1945.

Wordsworth, Dorothy. "Thoughts on My Sick-Bed." *Romantic Circles*, 2018. https://romantic-circles.org/editions/poets/texts/thoughts_on_my_sickbed.html. Accessed November 6, 2022.

Wordsworth, William. "Daffodils." In *Romanticsim*, edited by Duncan Wu. Oxford: Blackwell, 1998. 383.

———. *Kendal and Windermere Railway – Two Letters Reprinted from the Morning Post*. Kendal: Braithwaite and Son, 1845.

———. "Preface to Lyrical Ballads." In *Romanticsim*, edited by Duncan Wu. Oxford: Blackwell, 1998. 357–63.

———. "The World Is Too Much with Us." In *Romanticism*, edited by Duncan Wu. Oxford: Blackwell, 1998. 372.

Wordsworth, William, and Dorothy Wordsworth. *The Letters of William and Dorothy Wordsworth, The Later Years, Part I*. Oxford: Clarendon Press, 1967.

———. *The Letters of William and Dorothy Wordsworth, The Later Years, Part II*. Oxford: Clarendon Press, 1967.

Wu, Duncan. *Romantic Women Poets*. Oxford: Blackwell, 1997.

———. *Romanticism*. Oxford: Blackwell, 1998.

Yoshikawa, Saeko. *William Wordsworth and Modern Travel – Railways, Motorcars and the Lake District, 1830–1940*. Liverpool: Liverpool University Press, 2020.

Zilberstein, Anya. *A Temperate Empire*. Oxford: Oxford University Press, 2016.